THE
CREATION
SCIENCE
CONTROVERSY

THE CREATION SCIENCE CONTROVERSY

BARRY PRICE

MILLENNIUM BOOKS
Sydney • Philadelphia

First published 1990 by
Millennium Books Pty Ltd
3/32-72 Alice Street Newtown NSW 2042 Australia

National Library of Australia
Cataloguing-in-Publication data
Price, Barry.
   The Creation Science Controversy.

   Rev. and expanded ed.
   Bibliography.
   ISBN 0 85574 889 3.

1. Creationism. 2. Evolution-Religious aspects.
I. Price, Barry. Bumbling stumbling crumbling theory of
creation science. II. Title. III. Title: The bumbling
stumbling crumbling theory of creation science.

231.7'65

Designed by Luc Oechslin, Tatum Graphics, Sydney
Typeset in 11/13 pt Cheltenham by Printed Art, Sydney
Printed in Australia by Australian Print Group, Maryborough.

Contents

Acknowledgments

This book is the end product of several other books over the years and it is proper to acknowledge the support given by many people. In 1984 I was concerned that many teachers in Catholic schools were uncertain whether Catholics could believe in evolution. Subsequently, *Genesis, Evolution, and Creationism* was written and sent to schools with the support of Brother W. X. Simmons, then director of schools for the northern region of the Catholic Education Office of Sydney. Archbishop (now Cardinal) E. J. Clancy kindly gave permission to publish excerpts from his works on the origins of Genesis. In 1986 when I moved to the Sydney office of the CEO he supported me when maliciously attacked by Catholic fundamentalists in the church media.

In June 1986 Brother Simmons asked me to write a series of position papers for the CEO. The first of these, *The Two Books of God*, published in October 1986, gave religious reasons why Catholic schools could not teach creationism. It received wide publicity and is still the strongest statement on the issue from within Catholic education in Australia. Catholic schools owe a debt to Brother Simmons for his courage in confronting the powerful forces of Catholic fundamentalism.

This was followed in November 1987 by a booklet written for the nonscientist, which exposed the scientific nonsense of creationism, thus complementing *The Two Books of God*. A furor of publicity followed the publication of *The Bumbling, Stumbling, Crumbling Theory of Creation Science*. It displayed clearly to the average man in the street that creation science is utter nonsense. Less than two months after its release, publication was suspended by the CEO on orders from the Archdiocese. The reason given was that legal action had been taken against it and the matter was in the hands of the Archdiocesan lawyers. Six months later the ban was made permanent.

In the meantime some of the 2000 copies of *The Bumbling,*

Stumbling, Crumbling Theory of Creation Science had found their way to the United States. When I left the CEO at the end of 1987 I was given all rights and subsequently gave permission to a number of people in America to copy it at will. It seems that there are now more copies in the United States than in Australia. A number of American colleges are using it as a student text or a resource in courses on creationism.

While the opinions expressed are my own I would like to thank again the people acknowledged in that book: Dr. Alex Ritchie and Mr. Ronald Strahan of the Australian Museum (Sydney), Professor Ian Plimer of the University of Newcastle, Professor Ted Fackerell of the University of Sydney, Dr. Bob Selkirk of Macquarie University and Professor Michael Archer of the University of New South Wales.

But there are two people to whom special thanks are due for their assistance in this present publication: Mrs. Marian Finger and Professor Thomas H. Jukes. Marian Finger is the mother of one of the children subjected at age 12 to creation science in a public school at Livermore, California. The teacher had appropriated federal funds for gifted children and the case subsequently became a nationwide scandal. Marian, a suburban housewife, was at the center of it. Her courage has been an inspiration and her more than generous help in supplying materials, checking drafts and guiding me through the intricacies of the American system has made possible the significant section on creationism in the classroom. My hope, Marian, is that this book repays some of the support and time you have given me.

There are many in America who have sent me materials unsolicited while offering assistance in any way they could. However, Thomas H. Jukes, professor of biophysics at the University of California, Berkeley, first made known overseas my booklet *The Bumbling, Stumbling, Crumbling Theory of Creation Science*. Tom is not only a practicing scientist, he has an international reputation for his writings exposing the nonsense called in Australia creation science and in the United States scientific creationism. Our two minds are as one in that we both feel strongly that the lifelong bigotry and prejudice inflicted on the young of both our countries is the most important area to counteract. We both agree that it is in our schools and universities that we are failing to counteract creationism. I thank you, Tom, for your unfailing support, your precious time and the volumes of material as well as the fruits of your experience and wisdom.

All the above mentioned made me decide not just to write another book about creationism but to forge a weapon against it. To place it in the context of its causes and effects arising not only from science but from religion and society. One cannot legislate against those who have yet to learn that minds, like parachutes, work only when they are open. Nor can one convince them. At the most, one can only hope that casting the light of understanding will help parents become aware of the darkness creeping into our schools. This makes the effort worthwhile.

Finally, I would like to thank my family for putting up with me over the past twelve months. Stephen Jay Gould recounts that while he studied creationist materials in order to write his books he was locked away downstairs in his study and totally unapproachable. The only sounds heard by his family were occasional deep bellows of BLOWWWW! I can verify that, although the expletives may have been less delicate.

Barry Price

Grateful acknowledgment is made to the following for the use of copyrighted material:

Creation Life Publishers for quotations from Duane T. Gish, *Dinosaurs: Those Terrible Lizards,* 1977; and from Gary Parker, *Dry Bones... and other Fossils,* 1979.

Baker Book House, for quotations from Henry M. Morris, *The Twilight of Evolution,* 1978.

Bethany House Publishers, for quotations from Henry M. Morris, *Remarkable Birth of PLanet Earth,* 1972.

Introduction to Creation Science

Creationism is an organized and well-financed attempt to replace science with palpable fabrications. Creationism is difficult to defeat because its exponents allege that their cause has a religious basis, and thus they gain public sympathy. Many scientists, especially evolutionists, waste much time in debating and arguing with creationists, who can manufacture falsehoods much more rapidly than these can be exposed by scientific arguments, which are inevitably laborious. This imbalance of fiction and fact is particularly difficult to cope with on television, where time is allotted in seconds.

Creationism had its beginnings with the publication of Charles Darwin's *Origin Of Species*. Darwin reasoned that all living organisms had a common evolutionary origin. This was a blow to anthropocentric beliefs, but most religious faiths soon adjusted to evolution, and accepted it as a manifestation of divine powers. However, a few so-called fundamentalist sects rejected it, and were particularly hostile to the conclusion that apes and human beings have a close evolutionary relationship. The hostility persists, in the face of overwhelming scientific evidence for this conclusion, which creationists consider to be demeaning. Creationists cannot accept evolution, because they are committed to an unchangeable dogma based on their literal belief in the Book of Genesis, as set forth in the King James version of the Bible. They interpret and embellish this literal belief far beyond the text of the Bible. Their beliefs, to which creationists are fully entitled, would be harmless were it not that they seek to impose them on all individuals, and to pervert the teaching of science to young people by a travesty called "creation science." Simultaneously, creationists attack evolution as being sinful and diabolical, and they would, if successful, eliminate the teaching of it.

In recent years, the growth of knowledge has provided a comprehensive view of the universe and of the place in it of our planet and ourselves. Astronomy, geology and biology enable us to estimate an age of 10 billion to 20 billion years for the universe, of 4.6 billion years for the earth, and about 3.5 billion years for terrestrial life. Molecular biology shows us that all living organisms have a common ancestor and that the periods of divergence of various classes (such as kingdoms, phyla, families and genera) of living organisms are measurable through the molecular evolutionary clock. In the meantime, the ideas of creationism have not changed. The declamations of William Jennings Bryan in the 1920s are parroted by Henry Morris in the 1980s. One thing has altered: creationists have become more cunning and deceptive. Barry Price has explained this in his book. They have evolved into using "big business" procedures, such as mass media and persuasion. They plead for fair play and "equal time," but when challenged, they retreat behind a facade of religion, or they claim that the speed of light has changed, or that inconsistencies can be explained by invoking miracles.

Creationism is an affront to common sense and a menace to truth. It must exposed, discredited and eliminated. Barry Price has accepted the challenge of so doing.

Thomas H. Jukes
Professor of Biophysics
University of California, Berkeley

Chapter 1

An Overview of Creation Science

This book is about a unique form of religious fundamentalism called "creation science" or "creationism" in Australia but more usually called "scientific creationism" in the United States because many creationists do not subscribe to creation science. There are in the world today a great many fundamentalist sects but creation science is unusual in the way it has been the subject of litigation in the U.S. Supreme Court. The plaintiffs were leaders from the major Jewish and Christian faiths as well as scientists. Creation science is attacked by scientists as a pseudo-science which uses fraud and deceit and lies. It is simultaneously accused by religious leaders of making a mockery of how the Bible should be interpreted. This is because creationists use the Bible as a science textbook.

In the United States, copies of the creationist book *The Twilight of Evolution,* by Henry Morris, were distributed free by evangelist Jerry Falwell as part of the electoral campaign of presidential candidate Pat Robertson. Although he never mentioned it again, Ronald Reagan in his 1980 election campaign stated that the biblical account of creation should be taught in schools as well as evolution. With 10 percent of the U.S. population acknowledged as creationist, the movement has the potential to make or break governments and presidential candidates. In parts of Australia the situation may already have been reached where some politicians perceive the creationist vote as necessary to maintain their continued presence in Parliament.

It is the near unanimous opinion of scientists that creation science is utter nonsense. It is attacked in a statement signed by seventy-two of America's Nobel Prize winners. It is condemned as of little scientific worth by both the Australian Academy of Science and the American National Academy of Sciences as well as seventeen state academies of science and the 28,000 elected members, nearly all with doctoral degrees, of the Federation of American Societies for Experimental Biology. Yet none of the above has been effective in halting its growth either in the United States or in Australia.

Creation science reached the shores of Australia in 1980 and has thrived since then. At times it poses as science, while at other times, as the need arises, it poses as religion. The major target of creation science is the schools of Australia. It claims the right to share equal time in science classes with the teaching of evolution. It has achieved this by both direct and indirect means.

Since its Australian beginnings, it has been approved by the Queensland Minister for Education and banned by the New South Wales Minister for Education. It has been banned from Sydney Catholic schools and strongly discouraged in Queensland Catholic schools. The South Australian Minister for Education has banned it from being taught alongside evolution but allowed creationist speakers and literature into government schools at the discretion of the principal. In the private Christian schools throughout Australia, which have mushroomed in recent years, it is both accepted and welcomed as a substitute for the teaching of evolution. It is also widespread in universities and other tertiary institutes.

In the years ahead, creationism will be certain to make the headlines more and more often in Australia. There also exists a strong possibility that, as in America, creationism will become an issue in the courts of Australia.

AN OVERVIEW OF CREATION SCIENCE

For many decades Australians have been aware of the various religious groups, mainly American in origin, who travel from door to door with a message of salvation. One may admire their dedication but one is free to accept or reject their message. There are many people who have found meaning in their lives with what these groups have to offer.

In more recent years accusations of "mind control" have surfaced against some groups, scientology for example. The Jonesville massacre in 1978 is the prime example of a religious cult gone wrong. Media coverage from the United States has indicated to Australian audiences the marketing genius of a few unscrupulous evangelists who can attract thousands of followers and extract millions of dollars from them. A common ingredient of such groups is a sense of urgency because the end of the world is near, coupled with an offer of salvation which, if refused, will mean going to hell.

In an era of mass communications and unprecedented social and cultural change, the best way to attract followers is media coverage.

As will be shown, creation science, since 1980 a more recent American import onto the Australian scene, is a *unique and brilliant marketing approach* for a particular brand of religious fundamentalism. It is targeted at schools and it is too serious a threat to the education of Australian youth to be ignored.

At first regarded as a "bit of a joke," it has spread into New South Wales from Queensland to such an extent that in 1987 some 20 percent of the first-year biology students at both the University of Sydney and the University of New South Wales stated that they were creationists who did not believe in evolution.

Toward the end of 1987 the legendary Sir Joh Bjelke-Petersen's rule as Premier of Queensland ended. For the first time in nearly two decades Queensland had a premier who was not a creationist. Many felt that this would bring about a change in the policy of government support for creation science in schools. But their hopes were soon dashed.

The new Minister for Education, Mr. Littleproud, stated in his first press release:

> A policy of acknowledging alternative theories on the origins of man will continue in Queensland schools ... We will continue to ask science teachers to provide balance in discussion on the theory of evolution and on alternative theories (*Sydney Morning Herald*, 10 December 1987, p. 15)

On 7 January 1988 the Brisbane *Courier Mail* reported on page 3 that the New South Wales government, under new regulations, "will have the power to close schools which taught the biblical story of creation rather than evolution in science classes." These new regulations apply to all nongovernment schools. Creationism had already been banned, in 1987, from science classes in New South Wales government schools. This ban had been supported by Catholic education authorities.

The *Courier Mail,* in the same article, reported that the new rules were bitterly opposed by New South Wales' thirty-five Christian community schools and a similar number of parental schools.

Prior to this, on 2 October 1986, Brother W. X. Simmons, director of schools for the Archdiocese of Sydney, sent to each Catholic school in the Archdiocese a document called *The Two Books of God*, in which creation science was rejected on religious grounds. In the Foreword, Simmons stated:

> The present publication, *The Two Books of God*, explores the basic relationship between nature and scripture. Fundamentalism, however, distorts and misinterprets this relationship. Scientists do not accept creationism as science, and the Catholic student who is taught creationism as if it were a science is forced into an impossible position. No professional teacher should consider doing such. (Price, 1986b)

Creationism and Evolution

Creation science sees evolution as the work of the devil. Braswell Dean, an American judge in the State of Georgia, summed up the creationist attitude to evolution:

This monkey mythology of Darwin is the cause of permissiveness, promiscuity, pills, prophylactics, perversions, pregnancies, abortions, pornotherapy, pollution, poisoning and proliferation of crimes of all types. (*Time,* 16 March 1981)

The creationist argument which is used over and over again is this:

X is bad.
X did not exist before Evolution.
Therefore, Evolution caused X.
Conclusions: (a) In order to destroy X, we must destroy Evolution.
                (b) X is evil and the work of the devil. Therefore, Evolution is evil and the work of the devil.

Creationists view evolution as the work of the devil and, in addition, label those who believe in evolution as under the influence of Satan. Henry Morris, the founder of creation science, states that

Consequently, until Satan himself is destroyed, we have no hope of defeating the theory of evolution, except intermittently and locally and temporarily ... But we can speak confidently of the imminent death of evolution because we can discern ample signs of the imminent "coming" of the Lord, which the latter-day scoffers so vigorously resist. The very fact that uniformitarian and evolutionary thought seems to have captured the intellectual world is noted by the Apostle Peter (II Peter 3:3, 4) as indicative of the "last days". (Morris, 1978, p. 93)

You may well say: So what? *Or*: Why be bothered trying to understand a group of religious nuts? *Or*: They've always been around claiming the world is going to end any time now, that anyone who doesn't listen to their teachings will spend eternity in the fires of hell.

Such comments would be an understandable reaction. But you would be quite wrong to dismiss the strength of *this* particular group of religious nuts. The Creation Science Foundation Limited, based in Queensland, has access to and uses the considerable expertise, materials, strategies and human resources of its counterpart in San Diego, the Institute for Creation Research.

Perhaps the above sounds a bit far-fetched. But before you reject these opinions out of hand, let's see what two Australian scientists have to say. Both are involved with teaching the approximately 20 percent of first-year undergraduates who have been led to believe the universe was created 6000 years ago.

Michael Archer, a professor at the University of New South Wales, claims that during two years of reading everything available in the creation literature, he risked "a self-inflicted frontal lobotomy." He has these comments on the geology of creation science:

To teach what "scientific" creationists want taught will mean telling students among other things: that the Earth (as well as the whole Universe) is only about 6,000 years old; that the vast majority of the fossil record formed in one year from a flood that occurred 4,000 years ago; that all dinosaurs, lions, snakes and venus fly traps ate plants until Adam sinned in the Garden of Eden (prior to which time there was no death); that after the flood receded 3,999 years ago, all kinds of organisms on Earth landed on a volcano in Turkey and walked (slithered or whatever) from there to every continent on the then flood-devastated Earth without stopping along the way in any inappropriate places (which explains, for example, why there are now no Platypuses in Turkey); that the whole of continental drift including the breakup of Pangaea and Gondwanaland took place in the last 4,000 years; that dinosaurs survived the flood 4,000 years ago by being on the ark and persisted to become the legendary fire-breathing dragons we read about in story books; and so forth *ad absurditum*. This is the "science" that Creation "scientists" want taught in science classes. (Archer, 1987a, p. 161)

Ian Plimer, professor and head of the geology department at the University of Newcastle, expresses the same sentiment:

I have no faith in creationism. Creationism makes a muddle about the beginning and thrives on lies, false logic, misquotes, distortions, double standards and heresy. The Creation Science Foundation is a supporter of unsubstantiated fringe theories, is unscientific, is not supported by any eminent scientist, is un-Christian, is intolerant, is racial, is political, is theologically damaging, is against truth, is against knowledge and is unnecessary. It is analogous to a fundamentalist Iranian muslim flat earth society enforcing its views onto our education system to bring all back to the Dark Ages.
Creationism is not a joke. It has the potential to dramatically erode our already declining standards of literacy and numeracy. As educated geoscientists, we cannot ignore the threat and must be involved in the education of our children, must lobby politicians (after all, the only good politician is a frightened politician) and write to newspapers ...
Why have all the creationists ignored the possibility that God created evolution? (Plimer, 1986, p. 7)

The Making of the World

Humanity enters the third millennium with a world view as different as could be from that held for tens of thousands of years. The world view today is that of a dynamic, evolving universe of vast size and unimaginable age. This is in stark contrast to the eons-old human perception of the universe as static and unchanging, with the earth at its center. This new world view is arguably mankind's greatest intellectual achievement.

Astronomy, physics, chemistry and radiometric dating

independently assign an age to the universe of between 10 billion and 20 billion years. The evidence is overwhelming when considered as a consensus from the many branches of science.

The age of the earth is currently estimated at 4.5 billion years, using radiometric dating. As far back as 1907 it was measured as older than 2.2 billion years, while in the 1860s it was estimated at around 100 million years (Ritchie, 1987, p. 58).

The analysis of amino-acid sequences in cytochrome-C (and other proteins) is probably the most compelling evidence of how life has evolved over the past 2 billion or 3 billion years from simple cells to the vast variety of life-forms which surround us today. This is quite apart from well-known evidence such as comparative anatomy and the fossil record.

Nor does any of the above depend upon the theories of Charles Darwin, with which evolution is popularly associated.

The opposite is true. More recent scientific insights indicate that neo-Darwinism is at best a partial explanation of *how* biological evolution occurs. The demise of Darwinian theory as *full* explanation in no way alters the firm consensus of science that the universe has evolved. There is at the moment not one competing theory which can account for the observed facts.

Nor has any evidence emerged from any branch of science to contradict the view that the world has evolved.

To say that there is a *complete consensus among scientists that evolution has occurred* does not mean there is complete understanding of the underlying mechanisms, or ways, in which evolution has occurred. Far from it. While evolution is a fact, how it occurs will always be the subject of debate. This is the fascination of science.

To put it another way, there is no dispute about the *fact* that evolution has occurred but there is dispute among scientists about *how* it has occurred.

Evolution answers some questions but reveals many more questions. Some of these questions at this stage appear to be unanswerable in the light of present scientific knowledge. In common parlance: "The more you know, the more you know you don't know.

There is no conflict nowadays between belief in evolution and belief in God.

Let Creationism Speak

Before we proceed to examine the range of creationist literature in detail it may be illuminating to examine briefly three examples in the form of comic books. Millions of these have been sold, although they are not listed in recent catalogues of the Creation Science Foundation. A mine of information concerning creationist beliefs and methods is contained in these tiny books.

In addition, an analysis of these three little books shows clearly the distinction between the squeaky-clean public image put forward by creationists of themselves as dedicated scientists battling the narrow-minded establishment. Also shown are underlying religious beliefs of creation science which are not made nearly as public. Both public image and private beliefs are derived from a literal interpretation of the Bible.

The first booklet – *Have You Been Brainwashed?* – is by internationally known Duane Gish, vice president of the Institute for Creation Research in San Diego. This is the "mother house" of Australian creationism.

In March 1988 he visited Australia and while in Sydney he engaged in a public debate with Australian scientists, which was later shown on national television. Gish is regarded as the premier debater of all creationism as well as being the author of a large number of books on the subject.

This booklet, comprising about fifty frames, is based on Gish's book *Evolution? The Fossils Say No*. The storyline is Gish lecturing to a public audience. He proceeds frame by frame to itemize all his reasons why evolution cannot be true. It is very convincing to the nonscientist. The implied message is that scientists are a bunch of closed minds who can't see the forest for the trees. While the booklet is replete with misquotes and errors, this point is not of concern to us at the moment.

The high point of this comic book is the inside back cover. Here Gish appeals to the reader to turn to God and read the Bible, the implication being *after* the reader has abandoned all evolutionary notions. The punchline is a warning from Gish: "All who refuse to accept God's love and forgiveness will receive terrible punishment."

Gish's booklet illustrates an ever present theme which permeates creationism:

YOU CANNOT BELIEVE BOTH IN GOD AND IN EVOLUTION.

This perennial theme of creation science can be written in the form of a logical syllogism:

Evolution is the work of the devil.
To choose the devil means going to hell.
Hence all who choose evolution go to hell.

The logic is flawless and the conclusion correct. But *only* if evolution really is the work of the devil. This explains the never ending litany of "evolution is the work of Satan" in creationist writing.

If repeated often enough and if the lie is big enough, belief will follow. The Nazi propaganda machine ably demonstrated this.

Thus the reader, particularly the young reader, is placed in a dilemma of choosing either faith or science. This dilemma where none exists is the projection of a mind fearful of its own faith.

The truth of the matter is that some find in their own experience a "god" whether it be found in nature, a relationship, evolution, or whatever. Others do not. It is as simple as that. Many scientists are unbelievers, while many are believers. This is true also of the general population. Using fear and guilt tactics to dominate others in the name of saving them is a sure sign of self-doubt in the area of faith. Moreover it is an intolerant arrogance which assumes that the creationists alone have all the answers about God. A god which can be contained in the mind of man is a puny god indeed. It is setting oneself up as a god.

However, one thing is sure. Opposition to creationism has united both believer and unbeliever, at least in the scientific community.

Big Daddy? is another miniature comic book, by Jack T. Chick. It has been shown, by Zindler (1985b, p. 28), that a substantial part of this booklet has been "lifted" from a Time/Life book titled *Early Man* , and the labeling significantly altered.

This deceptive booklet contains some atrocious scientific errors but is probably the most widely distributed of any piece of antievolution literature.

The storyline is similar to that of Gish's *Have You Been Brainwashed?* but the setting is a classroom in which a university lecturer is teaching evolution and one student protests and begins to question the lecturer. The result is inevitable: the lecturer rejects evolution and resigns from the university. The final two pages call for repentance and the reader is assured that if this does not ensue then "eternal loss of heaven" will follow.

This is one of the less subtle examples of creationist literature but probably the most effective ever produced. The message is the same:

A BELIEF IN GOD AND A BELIEF IN EVOLUTION ARE TOTALLY INCOMPATIBLE.

Big Daddy? contains what must surely be a prize candidate for the craziest explanations and the most far-out misuse of the Bible in creationist literature.

The finale for the unfortunate lecturer occurs when the student asks him what holds the two positively charged protons together in the nucleus of an atom, since positive charges always repel each other. The lecturer replies that he does not know. Upon hearing this, the student whips out his Bible and quotes: "by Him all things are held together" (Colossians 1:17). This converts the lecturer and he resigns because evolution "can't possibly be true."

The biblical quote may have a religious meaning concerning Christ and the world but it is hardly intended to be applied at random when two things stick together, whether they be protons, a cork in a bottle or a jammed window.

The Beast, also by Jack T. Chick, is extremist by any standard. The message is the same – believe or else – but this booklet is concerned with attacking the World Council of Churches which, it is claimed, has been infiltrated by Communist agents of Satan. Also agents of Satan are all those pastors and theologians who have discontinued teaching the Bible as literal truth. Such teachings are predicted in the Bible as a sign of the end times. The churches (Protestant only; Catholics don't rate a mention, so presumably they were damned long ago) have been taken over by Satan. The next event is the raising of all "good" Christians bodily into heaven. The first bodies to go will be those buried in cemeteries. Undisturbed graves will indicate those who must remain behind because they died in sin.

The next to go up will be the living who are true believers. They will go into a rapture and disappear from the face of the earth. Some creationists believe this is already happening. (This writer bought from a Christian bookstore a car sticker which read BE WARNED. THE DRIVER OF THIS CAR IS LIABLE TO DISAPPEAR AT ANY MOMENT. Placed on the passenger dash it led to many fascinating conversations!)

The booklet goes on to describe how Russia attacks Israel and is miraculously defeated; the Antichrist becomes a world hero and all the churches join with Satan (true believers at this stage have all disappeared into heaven).

After this there is a pseudo-peace and Satan controls the world through computers; there is a world conflict in the Middle East and plagues strike all mankind. The super-church is no longer of any use to the Beast and so is destroyed.

Christ comes again and after a great battle the unbelievers are thrown into a lake of fire. And so it goes. The message is that you, too, are doomed unless you believe in Christ – not just any Christ but the inhumane Christ they portray.

The illustrations are like scenes from a horror movie. While the booklet does not attack evolution as such, the message complements that in *Big Daddy?* – all Christians who don't follow a literal reading of the Bible are employees of Satan since the Bible is the Word of God and God cannot lie.

The Beast, as many will know, is derived from a literal translation of the Book of the Revelation to John, which is the last book in the Christian Bible. Written in the time of Nero's persecution of Christians, it is an example of apocalyptic literature, a form of writing used in different parts of the Bible which has no counterpart in Western literature. Apocalyptic writing used codes known to the initiated. The message was always one of hope in times of persecution when hope was low.

The codes were in the form of colors, numbers, certain phrases and parts of animals. The writing was designed to confuse the persecutors if it fell into their hands. Such literature is analogous to the coded messages broadcast to Occupied Europe by the BBC during World War II. It is the same code used in the Book of Daniel when the Greek tyrants persecuted Israel, and elements of the code are found in various other parts of the Old and New Testaments.

Combined, the three booklets illustrate the spectrum of beliefs common to creationist literal reading of the Bible:

- Evolution is unscientific, impossible and the work of Satan.
- To believe in both God and evolution is not possible.
- Belief in evolution means going to hell.
- Evolution is responsible for most of the present-day evils in the world.
- The mainline Christian churches who do not preach the Bible as literally true are under the influence of Satan or communism or both.
- Communism is a disguise of Satan, as is the World Council of Churches.
- The end of the planet is due any moment now.
- Believe or go to hell.
- Justice, peace, oppression, poverty, the environment are of no concern and never rate a mention since the end times are already here.

All of the above views are equally valid if the Bible is read as literal truth. An important strategy of creation science has been to play down the mention of Satan in its public image, but it fully exposes itself in its writings. It also attempts to distance itself from too apparent an alliance with the extremist factions of creationism which publicly proclaim the last five of the views in the above list. This is quite a balancing act for creation scientists because in so doing, they are liable to be seen by fellow creationists as denying the literal truth of the Bible.

Is it possible that at some future time some creationists will be accused of heresy by their fellow creationists?

In the same way, it is the extremists of creationism who proclaim that the earth is flat and that the sun goes around the earth. But the differences between the various creationist groups are a matter of degree and not of kind. Both of the above are equally valid in terms of the Bible taken *literally*. Past history shows that the flat-earth theory and the earth-centered solar system were bitterly argued about by fundamentalists of bygone years.

This is the stuff that creationists demand equal time for in the science classes in Australian schools. It is what they have already achieved in many schools in the United States.

Not Here!

But if you think the above is confined to the United States, an article in the October 1987 issue of *Australian Presbyterian Life* will jolt you back to reality. The article, by Peter Hastie, should not, of course, be taken as representing the views of the editor of the journal, nor of the Presbyterian Church as a whole. But it is quoted as an example of the many fertile fields awaiting cultivation by the mind-changers of creation science.

Hastie's article is called "Genesis – The Devil's Playground." He begins by stating: "You've got to hand it to him: Satan is a brilliant strategist." He ends by concluding that "whenever the Church believes Genesis is a mythical or symbolic presentation of truth, it has fallen for the lie of Satan."

In between, Hastie implies that all modern biblical scholarship is the work of Satan. His argument in favor of literal Genesis is that it has "an air of realism about it due to its strong smell of 'facts'."

It is not the purpose here to oppose Hastie's arguments but rather to use them as an illustration of the type of attitude which is widespread in some churches around Australia. This attitude would welcome with open arms creation science speakers and materials in order to bolster their own inadequate evidence that evolution is the work of the devil. In so doing, they turn a blind eye to the findings of more than a century of scholarship which clearly indicate that the Bible is not a book of science.

Use of the Bible as a book of science is widespread. Roman Catholic fundamentalist groups, among them the Newman

Graduate Association, Catholics for Defense of the Faith and the Catholic Federation of Parents and Friends Associations, have been vocal in their efforts to have creation science alongside evolution in the science classrooms of Catholic schools. They still are. The first and last of these groups operate freely in Sydney. In addition, two bookstores in Sydney operated by fundamentalist groups are major sources of creation science books.

The Newman Graduate Association in April 1987 organized a symposium for visiting American creationist lecturer Gary Parker. He argued, of course, against evolution. On the same platform and acting as supporting speakers were a priest from a university residential college run by Opus Dei as well as a Sydney parish priest who is vice president of the Newman Graduate Association (*Catholic Weekly*, 15 April 1987, p. 6).

The March 1988 visit by Duane Gish was billed by the National Alliance for Christian Leadership, whose headquarters are in Canberra, as a "special bicentennial visit of noted American scientist, biochemist, and author Dr Duane Gish." The rest of the page lists Gish's professional qualifications and engagements. Nowhere on the page is creation science even mentioned. This letter of 2 March is signed by a member of the Alliance's national executive. Someone is being duped.

The same agenda featured "Church Meetings at appointed Catholic venues."

Once more, this evidence emphasizes how creation science has a ready-made network just waiting to be tapped. It also indicates its successful strategy of using a public image which gives the impression that it has nothing to do with religion. Even the most cursory examination shows that it has *everything* to do with religion.

It has been said that if all the errors and misquotes were removed from the writing of creation science, there would be nothing left. It may be added that if all the biblical misinterpretations were also removed, there would be no reason to attack evolution.

Chapter 2

Fundamentalism and Creation Science

The 1970s and 1980s have seen fundamentalism emerge as a worldwide socio-religious phenomenon. Although not a new phenomenon, fundamentalism is most marked in times of social and cultural change.

Fundamentalism is characterized by an underlying insecurity which feels threatened by change. It feeds on fear and anger and provides opportunities for revenge upon those perceived as agents of change. It is capable of manipulation by those who play upon the basic ingredients of insecurity, fear and anger.

> Most broadly speaking, fundamentalism is a historically recurring tendency within the Judeo–Christian-Muslim religious traditions that regularly erupts in reaction to cultural change. Psychological studies describe its strongest adherents as "authoritarian personalities": individuals who feel threatened in a world of conspiring evil forces, who think in simplistic and stereotypical terms and who are attracted to authoritarian and moralistic answers to their problems. (Arnold, 1987, p. 298)

Patrick Arnold also states that fundamentalism is always religiously divisive and marked by attempts to purge, persecute or oppress liberals, moderates and conservatives. He goes on to say that

In its most extreme form, this purge is accomplished through execution or massacre. Assassin, the Arabic word for fundamentalist, implies one method of eliminating leading members of the religious opposition. Killing and torture have marked the fundamentalist outbreak of the 1980s in the Middle East. (Arnold, 1987, p. 298)

He comments further on American fundamentalists:

In the United States fundamentalists seem content with driving the opposition out of the religious community through formal excommunication, firing or silencing. In the last decade, a number of rightward-drifting Protestant churches have expelled ministers and theologians who refused to abandon liberal or moderate principles. The Lutheran Missouri Synod, for example, forced out an entire theological faculty (Seminex) that would not accede to a new fundamentalist agenda including biblical literalism; a similar development is already well underway in elements of the Southern Baptist Convention. (Arnold, 1987, p. 298)

Fundamentalism in the Past

The Galileo case is a well-known example of the recurring tendency to which Arnold refers. Galileo published a paper in 1632 in which he claimed that the earth went around the sun. For this he was condemned by the Inquisition because he contravened the Bible, which said that the sun revolved around the earth. Galileo's belief was nothing new at the time, but his telescope was. Copernicus in 1530 had already described the sun as the center of the universe. A similar situation existed in the early seventeenth century. In 1605 Sir Francis Bacon, the first philosopher of science as we know it, and a devout Christian, argued against those Christians who used evidence from the Bible that the earth was flat. He attacked the

vanity of some of the moderns who have with extreme levity indulged so far as to attempt to found a system of natural philosophy (in our terms, science) on the first chapters of Genesis, on the Book of Job, and other parts of the sacred writings ... because from this unwholesome mixture of things human and divine there arises not only a fantastic science but also an heretical religion. (Frye, 1983, p. 201)

Bacon also warned against mingling and confusing the "book of God's Word in Scripture and the book of God's Works in nature." Bacon may have been influenced by John Calvin, who maintained that "It is only a literalist reading of Scriptures which requires a direct conflict with the findings of legitimate science."
The tradition of the "Two Books of God" is an ancient one.

In the early church, the great Saint Augustine in one of his sermons thus called upon his listeners to observe "the great book ... of created things. Look above you; look below you; read it, note it." (Frye, 1983, p. 199)

It is interesting to note that there are now creationist groups in the United States who espouse the flat-earth theory as well as an earth-centered solar system (the geocentric theory). These groups are more truly consistent in their literal interpretation of the Bible than those creationists who reject a flat earth and an earth-centered solar system while at the same time maintaining that the earth was created in six days of twenty-four hours.
A truly significant difference between the Christian fundamentalism of previous centuries and that of today is the labeling as agents of Satan the beliefs and organizations with which creationists disagree. Examples are evolution, theologians and the World Council of Churches, as well as just about everything else. Once upon a time this was confined mainly to witches, heretics and black cats.
This omnipresence of Satan is nowhere more evident than in the writings of Henry Morris, the founding father of creation science. He claims the markings on Mars and the craters on the moon are remnants of a mighty battle between Satan and good angels. Moreover, he claims that the outer planets and asteroids are inhabited by evil angels who visit earth disguised as UFOs.
This is somewhat reminiscent of the fourteenth century when the Black Death wiped out a third of the people in Europe. The

similarity is that then, as now, the end of the world seemed imminent for much of the population and the dominance of Satan was blamed. Then, as now, the fear of hell was a powerful weapon.

The Public Image of Creation Science

The table indicates the biblical origins of creation science. Shown as well are the main scientific beliefs rejected by creation science.

The Bible and the " Science " of Creation Science

Text	Creationist Interpretation	Consequently Rejected by Creationists
Genesis 1	6-day creation	Evolution, man related to other life-forms, molecular biology, neo-Darwinism, etc.
Genesis 2 and 3 Adam and Eve, the Garden	Humans created as perfect adults, immortal, no death or suffering	The second law of thermo - dynamics, etc. That the universe proceeded from the simple to the complex. (Creationists say the reverse has occurred.)
Biblical chronology	Universe created in 4000 B.C.	Age of universe as 15 billion years. Age of earth as 4.5 billion years. Astronomy. Nuclear physics. The speed of light. Radioactive dating. Archeology.
The Flood	Cause of all present landforms, mountains, oceans, fossils	Endemic species: different life-forms on separated continents. Fossil evidence. Geological time scales. Continental drift and most of the rest of geology
Tower of Babel	Cause of the different races and languages	Most of the archeological evidence concerning ancient civilizations.
Joshua	Sun goes around the earth	Geocentric theory of Copernicus and Kepler.
Biblical cosmology	Flat earth, celestial dome with floodgates	Circumnavigation of the globe first by Magellan and nowadays by thousands. Photos of the planet earth taken from outer space

Behind the Facade

The public image of creation science is that of a scientific alternative to evolution based on using the first eleven chapters of Genesis as a science textbook. But this public image is only the tip of the iceberg, and we know that the dangerous part of an iceberg is not the tip but that which lies beneath the surface. It is essential to look at some other parts of the Bible to understand the full import of creationism. The following are just a few of the many creationist interpretations of some verses of the Bible.

Bible Text	*Creationist Interpretations*
Ezekiel 39:4	Satan rules Russia. Russia attacks Israel.
2 Timothy 4:3 - 4	The World Council of Churches is controlled by Satan.
2 Peter 2:1 - 3	Theologians and biblical scholars work for Satan.
Revelation 13:17	Credit cards are the mark of Satan.
1 Thessalonians 4:16-18	The bodies of all true believers (i.e. creationists) will very soon be lifted into heaven while Israel defeats Russia in a nuclear war. This includes believers buried in cemeteries.
2 Timothy 3:1-5	Just about anything or anyone not "creationist" is an agent of Satan.
1 Timothy 2:11 - 14	The man is the head of the family. A woman's duty is unquestioning obedience.
Book of Job	A sign of God's favor is to be rich.
John 1:10, 5:22	All those not "born again" will go to hell.

Judaism Without the Prophets; Christianity Without Christ

The Old Testament is like a three-legged stool. With care, one can still sit on a chair with a missing leg but a three-legged stool with one leg missing is useless. Creationists have removed the Prophets from the Old Testament and thereby eradicated God's love and mercy.

Consider a few quotes from the Prophets which are never found in creationist writings.

> The kind of fasting I want is this: Remove the chains of oppression and the yoke of injustice, and let the oppressed go free. Share your food with the hungry and open your homes to the homeless poor. (Isaiah 58:6 - 7)

The Old Testament Prophets denounced the evils which they saw in contemporary society. For this they were despised and persecuted. The times have not changed. Modern prophets as often as not receive the same treatment. But underlying the denunciations of the Prophets was the ever present message that no matter what happened, God's love was constant and everlasting.

> Can a woman forget her own baby and not love the child she bore? Even if a mother should forget her child I will never forget you. I can never forget you! I have carved your name in the palms of my hands. (Isaiah 49:15 - 16)

Jesus quoted the Prophets very often and also the great prayer of the Jews, the Shemah:

> Israel remembers this. The Lord and the Lord alone is your God. Love the Lord your God with your heart, with all your soul and with all your strength. (Deuteronomy 6:4 - 5)

The biblical conception of sin was to miss the target! The great St. Augustine stated that all sin was due to a failure to love enough and that all sin could be cured by loving more.

In his lifetime Jesus Christ was seen first and foremost as a prophet. His teachings emphasized love not hate. The Pharisees asked Jesus which was the greatest commandment in the Law. He answered:

Love the Lord your God with all your heart, with all your soul and with all your mind. This is the greatest and most important commandment. The second most important commandment is like it. Love your neighbor as you love yourself. (Matthew 22:37 - 39)

The above is a key element in Christianity. John records this emphasis also:

And now I give you a new commandment: Love one another. As I have loved you, so you must love one another. If you have love for one another, then everyone will know that you are my disciple. (John 13:34 - 35)

None of the above seems to come through in creationist writings. Quite the opposite. The overall view boils down to this: God created a planet and placed man upon it so that God could check out whether man would go to heaven or hell. Who would want to believe in this sort of god? Christ becomes little more than the avenging arm of God the Enforcer.

This distortion of Christianity engenders fear and guilt and is capable of being used as a weapon of manipulation. This same distortion, for those who see beauty in creation and in the other, is a powerful reason to become an atheist.

Belief in Jesus as the Risen Christ may be a technical definition of Christianity but by itself it hardly passes the test of the Founder. There is no statistical evidence that Christians live better lives than people of other religions. There is no evidence that unbelievers, atheists and agnostics are depraved. Nor that they have worse moral codes than Christians. To say that God loves only Christians is blasphemy. There are no churches in heaven whether or not one believes in heaven.

The omissions from the Bible by creationists are of the same magnitude of error as their attempts to show that evolution is the work of Satan. Creation science is as much an offense against religion as it is against science.

Chapter 3

Hoaxes of Creation Science

Certitude is a state of mind. It may be based on an exhaustive examination of the evidence or it may be based on scanty evidence supported by intuition.

Most people have been certain about something or other at some point in their lives, and later were forced to acknowledge, to themselves or others, that they were wrong. We are often defensive about our certitudes and do not take kindly to others who disagree with them or are even so impolite as to call them nonsense. We do not like to admit we are wrong, nor do we like to admit doubt into those areas of our lives where we have constructed our certitudes. The truth of the matter is that we gradually learn that no matter how certain we are, we can always be wrong.

All of this philosophy is stated succinctly by the adage that there are only two things certain in life: death and taxes.

The history of the Christian West is littered with persecutions and religious wars in which both sides were certain that their version of some doctrine was right and their opponent's wrong. This is occurring right now in the Middle East.

Religious certitude can lead people to go looking for every little shred of evidence to bolster their belief, whether it be looking for miracles or concluding that finding a taxi in peak-hour traffic on a rainy night in the city is due to the direct action of God. This seems harmless enough in the main, although the person who just missed the taxi may not agree.

The dark side of religious certitude is also the most potent weapon to prevent critical appraisal of religious beliefs. It is to label those who disagree as agents of Satan. As we have seen, creationists are extremely fond of doing this but they have antecedents in the torturing of heretics and the burning of witches in not too far distant times. Then, as now, it was overlooked that the one finger of accusation which points to the other person is always accompanied by three fingers pointing back to the self. If Satan has a hidey-hole it is most likely to be in the hearts of those who find Satan in everyone but themselves.

Another side of all this is that religious certitude can sometimes cause extreme gullibility, even in the best of minds. This is especially true when Satan can be used whenever there are difficulties in explanation or when the facts contradict the evidence. Creation science uses both "god of the gaps" and "Satan of the gaps," whichever is more convenient at the time.

A Round Sun Orbits a Flat Earth

The premier performing artist in all of creationism is Duane Gish. He has been called El Supremo and Young Lochinvar. Gish is the most powerful influence in creation science today and it seems only a matter of time before he assumes the mantle of full control from his leader and mentor, Henry Morris.

The strategies used by creationists to combat evolution have contributed significantly to the success of creation science both in Australia and in the United States. Gish has very much influenced, and continues to influence, creation science in Australia. His writings, arguments and debating techniques serve as a model for creation science.

Gish travels widely around the debating circuit in the United States and is a frequent visitor to Australia and the United Kingdom. In the area of debating against evolutionists his record is formidable and he seems to wield an unconquerable sword.

More than anyone else he has been responsible for distancing creation science from the more eccentric elements in creationism and thereby attempting to present the public image of creation science as a group of scientists lured on by the search for truth. It is the plight of these scientists to be persecuted and ridiculed by the establishment because of the threat which creationism offers to established science. Even Gish would be forced to admit that 99 percent of the two million or so working scientists around the globe have never heard of him.

But does this idol have feet of clay? Frank Zindler seems to think so.

Zindler calls himself a "creation watcher." He is one of several Americans who travel around the country to attend creationist conferences the better to know them, analyze them and fight against them. Creation watchers are a continual embarrassment to creationists in the United States. Zindler, formerly a professor of biology and geology, is now a science writer. As with almost everyone who spends any length of time analyzing creationist writings and beliefs, he writes with a dark sense of humor to cover the shadows, despair and hopelessness which arise in the mind from too lengthy a period watching human beings succumb en masse to their own self-deceits. There is exacted a price from those who immerse themselves in the nonsense, frauds and deliberate lies inflicted on the gullible by a leadership which has found a foolproof way of exploiting for their own gain the scientific naivete and religious yearnings of millions.

In part of his report on the 1984 North Coast Bible Conference in Cleveland, Zindler comments on a paper delivered by Gerardus D. Bouw, who has a Ph.D. in astronomy and teaches computer science at Baldwin-Wallace College in Berea, Ohio. Bouw proceeded to give mathematical proofs that the sun goes around the earth, which seemed to Zindler to be a mishmash of Euclid and Einstein.

But of course, the calculus was mostly there to confound the mentally sluggish. The *real* proofs were biblical. After explaining that the Missouri Synod of the Lutheran Church had taught geocentricity in its astronomy texts right up into the 20s of the 20th century, he quoted Joshua 10:13: "And the sun stood still, and the moon stayed, until the people had avenged themselves upon their enemies. Is this not written in the book of Joshua? So the sun stood still in the midst of heaven, and hasted not to go down about a whole day." (Zindler, 1986 - 87, p. 33)

Zindler quotes Bouw's final clinching argument:

> if God cannot be taken literally when He writes of the *rising* of the *sun*, then how can one insist that He be taken literally when writing of the *rising* of the *Son* ?

Over the three days of the conference another four speakers presented "scientific proofs" that the sun went around the earth. The conclusion of the conference appeared to be that Missouri Lutherans were bringing back the geocentric doctrine. It is ironical that at the same time on the other side of the globe the ink was scarcely dry on the document signed by Pope John Paul II which reversed the decision of the Church that condemned Galileo.

Gish and his followers were present throughout the conference but said not a single word .

Zindler describes in detail another speaker at the same conference:

> The silence of *all* creationists when geocentrists were speaking is quite puzzling. Does silence mean tacit acceptance? Embarrassment? Or is it a case of honor among thieves: If you don't expose me, I won't expose you.
>
> Whatever the answer may be for the most of the Ptolemaic talks, I can say with assurance that embarrassment was the cause of *everyone's* silence when Marshall and Sandra Hall (authors of the widely distributed book, *The Truth: God or Evolution?*) got up *together* to give one talk. As the discourse bounced back and forth between husband and wife every minute or so, things began to unravel.
>
> Clearly enough, they had explained that the heliocentric theory was a "Satanic counterfeit," and they told of traveling to the biblical plain of Gideon (where Joshua had commanded the sun *and moon* to stand still) and receiving a revelation that the *moon* is the clue to it all.

Without telling us how long they played twenty-questions with god after receiving this clue, the Halls proceeded to prove that the sun goes around the earth. Marshall had hardly launched into his "proof" before his train of thought became derailed. He groped for words and stalled. He couldn't find a way to pass the ball to Sandra. Soon he was weeping openly, announcing that god "any minute now" was going to give him the right words.

God not getting involved quickly enough, however, Sandra got back into the show and told that they had watched an eclipse of the sun in which the moon's "shadow" had moved *the wrong way*! (She never made it clear when she was talking about the moon's blackened image viewed against the sun, and when she was talking of the eclipse shadow moving across the earth's surface.)

Hope springing up eternal, she took two styrofoam cups and tried to model the motions of the sun and moon during the eclipse. Marshall stopped crying and gave encouragement.

But alas! Within another minute both were hopelessly befuddled by the Satanic counterfeit. Not only could they not realize that when facing the sun their left hands had faced east, (and) that when turning their backs to the sun (and the audience) their left hands were pointing west, they also seemed to be unaware that the pinhole cameras commonly used to view eclipses also reverse left and right. When their time ran out, they could only announce that they had given everybody the key with which to unlock the treasure chest of astronomical knowledge, and they implored those with the experience in the subject to go for it. But not even the Ph.D. astronomer tried to bail them out. Not *one* of the Christian scientists present offered to "throw out the lifeline to save the sinking savants." (Zindler, 1986 - 87, p. 33)

It is apparent that Gish was, and still is, on the horns of a dilemma. If he says the sun does not go around the earth, he contradicts the Bible and alienates a large segment of believers in order to maintain a squeaky-clean image for himself and creation science. If Gish follows the literal Bible on which is based the whole structure of creation science, then he lays himself open to ridicule in the eyes of the world at large.

There is no evidence that Gish or his followers have made a positive statement on this issue, although Zindler claims that "Gish is rumored to follow the heliocentric theory."

Earlier conferences would have presented Gish with a similar problem but the flat-earth element of creationism seems to have been absent in recent years. Perhaps it was too much even for the geocentrics.

In June 1986 the National Creation Conference was held in Cleveland. Neither Gish nor his followers attended.

So we still do not know if Gish believes in a round sun circling a flat earth. Do you suppose it makes a square orbit?

Osgood's Orange, Doolan's Tree and Ham's Sin

Dr. A. J. M. Osgood, according to an undated Creation Science Foundation audio cassette catalogue received in 1987 together with a 1986 - 87 supplement, is a medical practitioner with years of medical practice and study of history which "leave him highly qualified to handle related subjects." Osgood is a director of the foundation as well as its specialist writer on biblical chronology.

Of approximately 200 tapes listed in the catalogue and the supplement, eighteen are ascribed to Osgood. Of these, eleven deal with his biblical chronology in which he follows Ussher's 1650 chronology, although at least one creationist writer considers that Osgood's use of a computer has significantly improved on the accuracy of Ussher's attempts in the seventeenth century. Three of the remaining seven tapes are concerned with how to overcome the presence of Satan, for example *Satan's Attack on the Biblical Family Blue-print*. According to Osgood, Satan seems to abide in most Australian homes. Apparently those families who do not believe that God made the world 6000 years ago, and that all living forms on earth descend from the survivors on Noah's Ark, are particularly vulnerable.

So much for Osgood's scientific background. Now for his orange. On page 11 of the December 1987 issue of *Creation ex Nihilo*, there is an article by Osgood with a half-page photograph of an orange labeled as "petrified." Osgood found this among a display of rocks in someone's backyard in Kingaroy, Queensland. It appears to be a fossil and he uses this orange to prove that fossils are formed very quickly. There is nothing in the way of proof except that the orange is hard. The crust has not been analyzed, nor has the orange been cut into sections. It could be coated with concrete or glue. The inside could still be edible. Yet creationists continue to complain that scientific journals won't publish their "research."

A perusal of creationist writings indicates that almost anything at all is published if the writer includes some statement which claims that the earth is less than 6000 years old.

This same "scientific method" is used on page 10 of the same issue of the same journal. Robert Doolan, secretary of the Creation Science Foundation, and Jenny Bergman, Ph.D., are joint authors of an article which describes California's redwood trees, famous as among the tallest and oldest trees in the world. They are a well-known tourist attraction. The main point of the article is that because not one of them is older than 4000 years, this is evidence for the Flood.

However, Doolan and Bergman should have extended their visit a little to the White Mountains, also in California. There they would have found species of *Pinus longeva* which are over 5000 years old, as measured by actual tree ring counts. This is a thousand years before the Flood.

The creationist research discussed above is in contrast to scientific research which is communicated by publication in thousands of journals each specializing in some area of science. They are available internationally. Before a paper is published it is carefully scrutinized by other scientists familiar with that area of science. Such journals become a permanent record and any find of significance is quickly examined and verified by a worldwide community of scientists. Woe betide any scientist who is found guilty of submitting fraudulent results. Such a person will never publish again and will most likely be seeking another livelihood.

With creation science research, almost anything goes. The criteria for publication in creation science journals seem to be nonexistent so long as the article attacks evolution in some way. To keep up appearances creationist journals are interspersed with neutral articles which describe scientific findings not related to evolution.

Ken Ham is a founding member of the Creation Science Foundation and is currently working with the Institute for Creation Research in California, the headquarters of Gish and Morris. He has worked on a set of six videos which are presumably intended to replace the *Origins* series. These were not made by creation scientists and are hence marketed on a commission basis only. His first book is titled *Evolution: The Lie* . To help ensure that the meaning of the title is not lost, most of the front cover is taken up by a very realistic picture of the head of a serpent in tones of green and black. Ham gets ten out of ten for making the medium into the message.

Ham fully subscribes to the creationist line that the first humans were created instantly as perfect adults who were also immortal. As everyone knows, Adam and Eve blew it all for the rest of us and Ham never lets them forget it.

In an article titled "Sin," under the apt heading of God's Reminders, Ham (1988, pp. 44 - 5) finds sin in places you would never dream of. He even finds reminders of sin when pruning the roses. This is because of Genesis 3:18: "Cursed is the ground because of you." These are the words spoken by God after Adam and Eve were forced to vacate the Garden.

The zoo in particular is a most powerful reminder of sin for Ham, because here there are meat-eating predators. This is because the Bible says that before the sin of Adam there was no death in the world and all animals ate leaves and grass. One wonders if Ham has a cat.

Visiting a hospital is an equally powerful way to be reminded of sin. Not because of sympathy with the suffering but because there was no death or disease before the first sin. In the same way, buying clothes is a reminder of sin, as is going to work. Some of us may agree with the last, particularly if suffering from the morning after the night before.

Surprisingly, Ham does not mention the navel as a reminder of sin. For John C. Whitcomb (1972, p. 12), the navel is very much a reminder of sin. He says that although Adam and Eve were "created full-grown (without navels)," nevertheless they were fully capable of being fruitful and able to multiply and fill the earth. One must conclude that if the first couple had not sinned, and had thereby passed on their immortality to their children, there would not even be standing room on the planet!

But perhaps Ham is more widely read than Whitcomb. Two years before Darwin published his *Origin of Species by Means of Natural Selection* in 1859, Philip Henry Gosse had published *Omphalos*, which is Greek for "belly button." In this book Gosse argued that Adam and Eve would surely have navels even though they had no mother. He disagreed with Sir Thomas Browne who denied Adam a navel because it would be superfluous (Gosse, 1857, p. 289).

Gosse asked his friend Charles Kingsley, a minister and the author of *Westward Ho* and other books of note, to write a review. He refused. Kingsley replied to Gosse in part:

> Shall I tell you the truth? It is best. Your book is the first that ever made me doubt (the doctrine of absolute creation), and I fear it will make hundreds do so. Your book tends to prove this – that if we accept the fact of absolute creation, God becomes God-the-Sometime-Deceiver. I do not mean merely in the case of fossils which pretend to be the bones of dead animals; but in ... your newly created Adam's navel, you make God tell a lie. It is not my reason, but my *conscience* which revolts here ... I cannot ... believe that God has written on the rocks one enormous and superfluous lie for all mankind.
>
> To this painful dilemma you have brought me, and will, I fear, bring hundreds. It will not make me throw away my Bible. I trust and hope. I know in whom I have believed, and can trust Him to bring my faith safe through this puzzle, as He has through others; but for the young I do fear. I would not for a thousand pounds put your book into my children's hands. (Hardin, 1984, pp. 164 - 5)

The great St. Augustine's view of sin comes as a breath of fresh air compared to that of creationists. Although he well recognized man's tendency to sin, St. Augustine, as we noted before, saw all sin as caused by a failure to love enough. His remedy for sin was to learn to love more.

Once again, the biblical view of sin in the Old Testament is simply to miss the mark!

Publications and Misquotes

If there has ever been a creationist article or book in the area of science or scientific research that does not contain errors or a possible element of deceit, then it rests unknown and unnoticed in some faraway place. This should not be surprising since their basic assumptions are in error. That is far from saying that such errors are obvious.

Creation science does not count among its followers one single scientist of international repute. Much of creationist "research" involves seeking out scientific articles, often by famous scientists, which are published as supporting the creationist point of view. This nearly always involves quoting out of context, or worse.

A few creationists are competent scientists but rarely do they write in their own specialty. Henry Morris is a hydrologist of repute but his creationist writings are in astronomy, atomic physics, thermodynamics and geology. That is, in almost every branch of science except hydrology. Morris also has to his name voluminous writings on the Flood which he sees as the keystone of creation science. Presumably it is in these writings that he has found an outlet for his scientific specialty. The creationist science cupboard is almost empty on the chemistry shelf and completely bare on the physics shelf.

The Long Search

> A recent study showed that of some 135,000 papers submitted to major scientific journals over a four-year period, only 18 were on creation science. Not one of these was accepted for publication. (Scott and Cole, 1985, p. 23)

Eugenie Scott and Henry Cole (1985) searched the scientific literature to find what scientific papers had been published by twenty-eight prominent creationists. Their search spanned nearly four years of publication. They found that only six of the twenty-eight had published anything at all and that 75 percent of the fifty-two publications had been written by two of the six writers. All of the articles were in areas such as food processing, simulation of loads and vibrations in aircraft wings. Not one of the articles "dealt with theoretical support for the assumptions and concepts upon which scientific creationism is based" (p. 23).

As the above-mentioned survey shows, only eighteen creationist articles were submitted to scientific journals in four years. Fifteen were rejected and three were still under review.

> Reviewer comments regarding rejected articles complained about poor presentation ("ramblings ..."; "no coherent arguments ..."; "high-school theme quality ..."; "tendentious essay not suitable for publication anywhere ..."; "more like a long letter than a referenced article"), and failure to follow accepted scientific canons ("no systematic treatment ..."; "does not define terms ..."; "flawed arguments ..."; "failure to acknowledge and use extensive literature on particular questions ..."). (Scott and Cole, 1985, p. 26)

Creationist Writings

Creationists like to say at every opportunity that scientific journals reject their articles. The reviewers' comments above went on to say that the eighteen articles seem to have been written by laymen with no scientific training. It is apparent from this that those creationists with scientific training do not submit articles on aspects of creation science. Nothing on Flood geology, the speed of light, creationist thermodynamics, fossils or the age of the earth, and so on. These scientists who espouse creationism know very well that creation science falls apart under critical analysis. The only place they can get away with such scientific nonsense is in their own books and journals which are written for the gullible, where there is no critical review and where a single edition may publish contradictory views from different writers as a matter of course. It would seem obvious that the sole criterion for publishing in creationist journals is "anything goes," provided it attacks evolution, looks convincing and has a few references to make it seem scientific.

There is, of course, much debate about *how* evolution has occurred but the *fact* that it has occurred is not a matter of contention. The scientific evidence for evolution *does not depend* on the theory of Darwin. The evidence for the fact of evolution is derived from different branches of science using different experimental methods. Darwinian theory is an attempt to explain the *way* in which biological evolution has occurred. It is not a sufficient explanation for the *mechanism* of evolution. The point is, not being able to fully understand how something has happened does not alter the fact that it *has* happened.

It would appear that, by quoting out of context, creationists seek to give the impression that scientists themselves have doubt about the *fact* rather than the mechanism. But creation science goes much further than quoting out of context.

Martin Bridgstock (1985, p. 31) followed up ten quotes selected at random from creationist literature which included *Creation ex Nihilo* and *Prayer News*. Both of these are published by the Creation Science Foundation of Queensland. He found that these ten quotes yielded twenty-two errors and distortions. Nine of the distortions comprised selecting material so that it contradicted the main gist of the article, with one quote "so grossly incorrect that it is hard to see it as an error." There were incorrect dates, minor misquotes, omissions and mis-descriptions in the other twelve errors. Only one of the ten quotes was free from error.

Scientists Respond

Stephen Jay Gould is a professor of geology at Harvard University, the author of *The Panda's Thumb* and probably the single most misquoted and misused scientist among the creationists' unwilling allies. He tells what he thinks about it all in this excerpt from "Evolution as Fact and Theory," *Discover*, May 1981:

It is infuriating to be quoted again and again by creation-ists – whether through design or stupidity, I don't know – as admitting that the fossil record includes no transitional forms. Transitional forms are generally lacking at the species level but are abundant between larger groups. The evolution from reptiles to mammals ... is well documented. Yet a pamphlet entitled "Harvard Scientists Agree Evolution Is a Hoax" states: "The facts of punctuated equilibrium, which Gould and Eldredge ... are forcing Darwinists to swallow fit the picture that (William Jennings) Bryan insisted on and which God has revealed to us in the Bible." (Cole, 1981, p. 38)

Transitional forms refer to the Darwinian theory which held that the transition between species, for example from dinosaurs to birds, was by a series of gradual changes. Until recently this was thought to be the mechanism by which evolution occurred.

David M. Raup is the dean of science at the Field Museum of Natural History in Chicago. He is a "punctuationalist" whose writings, along with those of Gould and Eldredge, are among the most influential contributions to that theory. They are often cited by creationists in an attempt to bolster the creationist view that each species was created separately by God without transitional forms.

Raup comments below on the punctuated equilibrium theory of biological evolution, that biological evolution did not occur gradually as claimed by Darwin. He makes the distinction between fact and mechanism which creationists refuse to accept.

One of the most unfortunate aspects of the current creation-evolution debate is that many of the creationists equate Darwinian theory with evolution. They are saying, in effect, that if Darwin's theory falls, then so does evolution. Nothing could be further from the truth. To me, there are two basic questions: Has evolution occurred (in the sense of change in the biological composition of the earth over millions of years)? By what mechanism has evolution occurred? Darwin's contribution was to the second question. He proposed a biological mechanism: natural selection. Whether Darwin was right or wrong has no bearing on the question of whether evolution did or did not occur.

On the question of whether or not evolution has occurred, I would say that there are few things in the natural sciences about which we can be more confident. The geologic time scale has been checked and rechecked by many independent methods. Although individual dates may be subject to error, the overall chronology stands firm. It is used every day in petroleum and mineral exploration, and, if there were basic problems with it, I am sure that industrial geologists would have blown the whistle. The fossil record is intimately tied in with this chronology and shows a record of change in organisms through time. What we are not sure about is just how the biological changes took place. Natural selection surely played a part, but there may be other biological processes that have operated. One of the challenges of biology and paleontology is to find out what other processes were involved. (Cole, 1981, p. 39)

In 1986 a U.S. Supreme Court action was begun concerning the State of Louisiana's action in legislating for equal time in schools for creation science and evolution. This triggered a document unprecedented in the history of science. One of the briefs against creationism in schools was on behalf of seventy-two of America's Nobel Prize winners. Each of the seventy-two individuals had received the Nobel Prize in either physics, chemistry, physiology or medicine. Joined with them in their brief as to why creation science should not be taught in schools were seventeen state academies of science and numerous other scientific organizations. A powerful voice.

The document, twenty-seven pages long, contains twenty-two pages of detailed argument against the science of creation science, showing that it is religion not science. One of the creationist writers whose name appears in the brief on behalf of the Nobelists was Australia's own Andrew Snelling.

Snelling is a geologist and full-time worker with the Creation Science Foundation in Brisbane. He has an honors degree and a doctorate in geology. There is no question about his professional competence. Snelling is "actively engaged part time on uranium geology and exploration for a mining company and various

Australian and overseas research organizations" (CSF cassette tape catalogue, undated). He is one of the very few creation scientists who writes and speaks for creation science in the field of expertise in which he is trained and qualified. This is where his problems begin.

Snelling is accused of gross contradictions in his own writings. At the same time that he is writing in scientific journals about rocks 2 billion years old, he is publishing, in creationist journals, articles about the same rocks that are only 6000 years old.

The CSF speak and write prolifically using a considerable budget. They have made no contribution to science, presented not one iota of new data, have presented no original information whatsoever to support creationism, have performed no experiments, have engaged in no historical research, have undertaken no dispassionate observation and yet claim to comprise eminent scholars. The only exception is A. A. Snelling (CSF, Brisbane) who published a scientific paper in 1985 (*Jour Geochem Exploration*) on radiogenic Pb isotopes and Proterozoic geology. The same Snelling argues in the CSF literature debunking radiometric dating and supporting a 4004BC creation age of the Earth I leave the matter of honesty to the reader. (Plimer, 1986, p. 7)

Snelling is also in trouble over his conclusions concerning the coal seams of the Hunter Valley north of Sydney.

Other enlightened new data by Snelling are reports in the CSF literature of the occurrence of fossil gold chains and iron anchors in Australian coal seams supporting the CSF concept that coal seams are young, were destroyed by catastrophic volcanic explosions and formed instantaneously.
I'm somewhat confused why we don't mine coal seams for gold. Furthermore, if gold was brought to Australia in 1788 and coal was first discovered in Newcastle in 1791, then the Great Flood and catastrophic volcanism occurred in

Australia between 1788 and 1791. I'm not surprised that Australia is the centre of catastrophism, however the run must have been so strong in 1788/91 that all inhabitants did not notice and yet survived these two catastrophes. The same author also "observes" that Permian coal logs have a brown coal core and a bituminous coal periphery as a result of sudden coalification after catastrophic volcanism. This is blatant cooking of the data! (Plimer, 1986, p. 7)

Ian Plimer, who is professor and head of the department of geology at the University of Newcastle, situated close to the Hunter coalfields north of Sydney, offered $20,000 to anyone finding a gold chain in a coal seam. There are some 25,000 miners working underground each day in the mines. Plimer still has his $20,000.

There is another side to the gold chains. If there were gold chains as well as anchors in the coalfields, they must have arrived with or after the First Fleet of convicts in 1788. However, creationists claim that all coal in the world was formed at the time of the Flood. Thus, according to this reckoning, in 1988 Australia was celebrating not only the bicentenary of the First Fleet but also the bicentenary of the Great Flood.

One must sympathize with Snelling. Besides coping with Plimer attacking him from the front he has been constantly called on by Alex Ritchie of the Australian Museum to debate in public his creationist writings on geology. He refuses. Neither has he published anything in creationist literature in the past few years, although this may change following his election in 1988 as one of the seven directors of the CSF. He subsequently became a full-time worker with the foundation as its expert in geology.

The August 1987 issue of *Prayer News*, a bimonthly newsletter which gives information on the CSF's activities, responded by devoting half of its space to a near libelous attack on Plimer which did everything but deny that Plimer's accusations against Snelling were true. There was not a single word by Snelling but Carl Wieland, head of the foundation, in the latter part of his article asked readers to pray for Snelling and to pray that Plimer's attacks will not cause too much damage in his relations with his employers.

What seems to upset Wieland most of all is that Plimer has not followed the traditional academic procedure of debating with the niceties of language and in a gentlemanly way, as Wieland claims creationists always do. He is certainly right on this point. Anyone who reads Plimer's article in the *Australian Geologist* cannot fail to notice that he has called a spade not just a spade but a very large shovel.

The moral for creation science writers is this: *Always write on something you know nothing about.* That way no one will notice. Everyone will assume it is just another run-of-the-mill creationist article.

Making a Monkey Out of a Man

To completely check the accuracy of all references in any one of the hundreds of thousands of scientific papers published in the hundreds of specialist scientific journals can be an extremely time-consuming task. Scientific journals communicate to all scientists in the world what each individual has achieved, thus preventing replication while at the same time providing a base on which others may build. Together with conferences they are probably the single most important avenues for the advancement of science.

To find the wrong page number or year in a reference is an annoyance, but it does occasionally happen – usually the result of a typesetting error. However, a mistake in calculation will be quickly noticed and this rarely happens. Each paper submitted to a particular journal is scrutinized by reviewers who check for accidental errors and make sure that the findings of the paper itself are of sufficient quality to warrant being published.

Deliberate fabrication of results or quoting another's work as one's own is the ultimate crime. It strikes at the very heart of the scientific enterprise. The penalty for deliberate fraud is instant "excommunication," even grounds for legal action. If it occurs in a doctoral thesis the penalty is instant failure and a life sentence. The unfortunate recreant may not apply to do a doctorate in any university in the world. This code of the worldwide scientific community is rigidly enforced.

In light of the above and what follows, it may be fortunate for Duane Gish that he already has his Ph.D.

What follows is a resume, with some direct quotes of Frank Zindler's painstaking analysis of a portion of a best-selling book by Gish, *Evolution? The Fossils Say No.* Zindler comments at the beginning:

When one reads the creationist literature, one quickly comes to see that deliberate distortion – not just misunderstanding of the facts – is a major characteristic of the genre. Furthermore, unlike real science, which is self-correcting and usually exposes its own hoaxes quickly, creation "science" either corrects its frauds not at all, or only under irresistible pressure from real science. Like cancer, creationist errors and distortions simply metastasize, becoming more widely distributed and more deeply implanted. (Zindler, 1985a, p. 23)

The background concerns the discovery on 2 December 1929 of the first fossil skull of *Sinanthropus,* which became known as Peking man and is still regarded by Chinese scientists as the "sensational event of prehistory in China" (Zhou and Li, 1983, p. 39). Associated with the discovery was Teilhard de Chardin, who spent twenty-two years in China and wrote 300 scientific papers of which 120 were on China. His contributions in particular to the geochronology and stratigraphic succession of the Tertiary and Quaternary periods in China are still regarded by present Chinese scientists as of importance.

Zindler was told that Marcellin Boule, a French anthropologist who died in 1942, had said he

examined the Peking man site and the fossils, (and said) the fossils were from a monkey that probably was killed for food. (Zindler, 1985a, p. 23)

Zindler was mystified: he knew Boule had never said anything like this. He then remembered that

Evolutionists who are experienced in combating the

distortions and delusions of creationism have an adage they use when creationist materials smell ranker than usual: *"Cherchez la Gish!"* – referring to Duane Gish, the premier performing artist of all creationism. (Zindler, 1985a, p. 23)

Zindler then searched through Gish's book, *Evolution? The Fossils Say No* , and found the key to the mystery in the following paragraph from page 129 of that book:

In an article published in 1937 in *L'Anthropologie* (p. 21), Boule wrote:
"To this fantastic hypothesis (of Abbe Breuil and Fr. Teilhard de Chardin), that the owners of the monkey-like skulls were the authors of the large-scale industry, I take the liberty of perferring (sic) an opinion more in conformity with the conclusions from my own studies, which is that the hunter (who battered the skulls) was a real man and that the cut stones, etc., were his handiwork (the nature of this stone industry will be discussed later).

In Boule's book (p. 21), there is nothing resembling the above quote taken from Gish's book (p. 129). At no place in Boule's book is there any suggestion that Peking man was monkey-like. Quite to the contrary, there is no mention of monkeys in the whole book. On *page 20* of Boule's book the following passage does occur:

To this hypothesis, as fantastic as it is ingenious, I may be permitted to prefer one which seems to me to be just (as) satisfactory, being simpler and more in conformity with the totality of what we know: the hunter was a true man, whose stone industry has been found and who made *Sinanthropus* his victim!

It is rather more revealing to take out Gish's words in brackets and read the passages in direct comparison. Boule's version and Gish's version, typesetting errors included, are reproduced *exactly* in the box below:

Boule's Original

To this hypothesis, as fantastic as it is ingenious, I may be permitted to prefer one which seems to me to be just a satisfactory, being simpler and more in conformity with the totality of what we know: the hunter was a true man, whose stone industry has been found and who made *Sinanthropus* his victim!

Gish's Version

To this fantastic hypothesis, that the owners of the monkey-like skulls were the authors of the large-scale industry, I take the liberty of perferring an opinion more in conformity with the conclusions from my own studies, which is that the hunter (who battered the skulls) was a real man and that the cut stones, etc., were his handiwork.

Even the most credulous person might have to swallow twice before accepting that a misquote such as shown here is accidental.

Seeing Is Not Believing

If creationists can wreak havoc and confusion with the written and spoken word, it follows that their films could well be considered masterpieces of deceit. When you search through a can of worms, it is hard to find the biggest worm but the first of the six-part series called *Origins* must surely be a prime candidate.

The *Origins* film series was released in 1983 and is a joint Dutch - American production. The first half-hour program is also named "Origins." The series has won awards from the Academy of Christian Cinematographic Arts and is professionally produced.

In a half hour the film shows that evolution contradicts the second law of thermodynamics and is therefore impossible. It is utterly convincing to a nonscience audience and even to those with some scientific background. The crunch point, as usual, is that evolution is the model approved by atheists while belief in creationism places one firmly on the side of God.

The star of the show is Professor A. E. Wilder-Smith who has an impressive list of degrees but has certainly never published his brand of thermodynamics in any reputable scientific journal. There are probably two million or so practicing scientists around the world, most of whom would have had to study the essentials of thermodynamics as part of their undergraduate degrees and many of whom use thermodynamics for part of their working lives. One can pause at the outset of the film to wonder why at least one of these scientists would not have seen that thermodynamics contradicts evolution. But this, of course, is not the case.

In this film Wilder-Smith says again what he has already said in his book, *The Natural Sciences Know Nothing of Evolution*. In a review of Wilder-Smith's book for the National Center for Science Education, Kenneth Christiansen concludes:

> In summary this book, while it is occasionally amusing, is science-trash. It does not belong in a science classroom. (Christiansen, 1984, p. 65)

Christiansen points out that Wilder-Smith's book is mostly concerned with the origin of life, an issue "which would be unaffected by any theory of evolution." He also states that the most fundamental flaw in the book is

> an apparent confusion or ignorance (it is hard to tell which) concerning our present understanding of the evolutionary process. (Christiansen, 1984, p. 65)

Wilder-Smith's version of the second law of thermodynamics in his book is the same as that given in the film. His thermodynamics

> deny the possibility of the growth of a tree or the formation of a galaxy, as well as the spontaneous chemical evolution of life. (Christiansen, 1984, p. 65)

But it is in his constant use of the "god of the gaps" that Wilder-Smith approaches most closely the scientific naivete of stone age man.

In the ancient cosmology of the Middle East the movement of the stars was a mystery. The biblical writers saw the stars, as well as the sun and the moon, as being moved by the finger of God (Isaiah 45:12). While this expressed their religious conviction that God was close in their lives, it is not considered as literal truth. In the same way, poets and mystics have always found, and even today find, the world alive with the living God. This experience may be a reality for many but it is outside the province of science. The "god of the gaps" is the attitude that anything in the world which cannot be understood requires the direct physical intervention of a god. In some religions this appears as a myriad of lesser gods concerned with the "housework" of creation while one supreme god is outside it all, aloof from these minor chores. This whole issue is a recurring theme in creationism. The Creator is reduced to a creator, one who is kept very busy running around fixing up all the little flaws that creationists find in creation.

Most of those concerned with the study of the origin of life from inanimate molecules see the "rules" as having been "encoded within the chemicals which made up the primative earth" (Christiansen, 1984, p. 64). Wilder-Smith's view would require direct physical action by a god every time a living cell came into being. This is occurring at an unimaginable rate in all living things on the planet! This would place God physically present in every atom, which is pantheism.

Theologians of the Middle Ages hammered out a rebuttal to the god of the gaps which still stands. It is stated as "the greater the distance between the agent and the act, the greater is the power of agent." Thus a person who picks up crumbs from the kitchen floor with fingers or dustpan is less powerful than one who uses a broom or vacuum. The very same argument applies to a god who directly created and a god who created through evolution.

This is not to say that either evolution or creation is "proof" for God. Attempting to prove or disprove God is futile. The decision to believe, or not believe, in a god belongs to the individual. It is an inalienable right which must be respected and which any democratic society must guard carefully. Whatever else it may be, it is a decision completely and forever outside the province of science.

Following upon Wilder-Smith's thermodynamics is a section where Isaac Asimov seems to espouse Wilder-Smith's creationist thermodynamics. Asimov is extremely convincing as he explains what is essentially an introduction to the second law (see Chapter 10).

The narrator continues and acknowledges Asimov's disagreement with Wilder-Smith but in a way that most viewers wouldn't even notice: "Scientists know of no exception to this law and, although Dr Asimov *does not personally recognize it as such,* many scientists believe that this is a very real and serious obstacle for evolution" (italics supplied).

Asimov is made to sound the odd man out. The implication of the narrator's statement is the opposite to that of the true situation. It is only a handful of creationists who see the second law as opposing evolution. In none of their writings do they indicate any understanding of the second law.

The credits at the end of the film make no acknowledgment of Asimov. Nor is there any indication that the segment is an excerpt from one of his many science films, taken out of context. The illusion is complete. The viewer is led to believe that Asimov appeared in person as part of a legitimate creationist production.

Alex Ritchie of the Australian Museum was extremely puzzled by Asimov's appearance in the film and his seeming to espouse creationist thermodynamics. Asimov's reply to Ritchie's letter of inquiry leaves no doubt as to what Asimov thinks about the matter. He wrote :

I have not seen, or heard of, the film you mention.
If I have been included it is, of course, without my
permission, and out of context.
I'm afraid that to sue them would be a frustrating job and
that the only thing I can do is to continue to write such
things about creationism as I can.
They quote me to show what a devil I am and then they
want to twist things to make me appear on their side. They
seem to be rather ambivalent about me.
In any case, I will just go about my business and pay no
attention to their yapping.

In an earlier and much briefer version of this book, published by the Catholic Education Office in Sydney in 1987, Asimov's letter was also reproduced to illustrate the deceitful nature of the "Origins" film. Subsequently the Creation Science Foundation claimed in the media to have permission (*Sydney Morning Herald,* 22 December 1987). A request to the foundation for a copy of the permission was refused except under the following conditions:

* in a court of law,
* in a public unedited face-to-face debate on national television, or
* in a similarly publicized debate on the second law in front of a university audience.

Hence the situation at present is:

* The CSF claims to have permission.
* Asimov claims it doesn't. * The CSF claims the Asimov clip is not used out of context.
* A group of extremely reputable scientists claim that it is used out of context.

Let the reader decide whom to believe. Whether yea or nay, the bottom line question is this: Why would creationists allow their most famous enemy to explain views directly opposed to theirs unless they were going to use it out of context? Why?

The Ark

The evidence that the first eleven chapters of Genesis are prehistory has been painstakingly accumulated by archeologists and biblical scholars over the past century. This work is still going on. Bernard Anderson describes how in 1872

George Smith, a young Assyriologist employed as an assistant in the British Museum, was sorting and classifying the tablets from Nineveh, when suddenly his eye was arrested by a familiar reference. "Commencing a steady search among these fragments," he wrote later, "I soon found half of a curious tablet which had evidently contained originally six columns ... On looking down the third column, my eye caught the statement that the ship rested on the mountains of Nizir, followed by the account of the sending forth of the dove, and its finding no resting place and returning. I saw at once that I had here discovered a portion at least of the Chaldean account of the Deluge." (Anderson, 1967, p. 16)

The story of Noah describes how men had become so wicked that God despaired of their ever changing. God searched the earth and commanded Noah to build a wooden boat because a flood of waters would be sent to destroy every living thing on earth except Noah, his family and the creatures aboard the boat. The Flood followed and only those aboard the Ark were saved (Genesis 7 and 8).
What is often overlooked is that the Flood story is one of reassurance and confirmation. After the Flood came the Covenant of the Rainbow, a sign – a promise – that never again would such an event occur.
God said to Noah and his sons, "I am now making my covenant with you and with your descendants, and with all living beings – all birds and all animals – everything that came out of the boat with you. With these words I make my covenant with you: I promise that never again will a flood destroy the earth. As a sign of this everlasting covenant which I am making with you and with all living beings, I am putting my bow in the clouds. It will be the sign of my covenant with the world. Whenever I cover the sky with clouds and the rainbow appears, I will remember my promise to you and to all animals that a flood will never again destroy all living beings. When the rainbow appears in the clouds, I will see it and remember the everlasting covenant between me and all living beings on earth. That is the sign of the promise which I am making to all living beings." (Genesis 9:8 - 17)
Perhaps even in those times the biblical writer felt a need to reassure people against prophets of doom wandering around trying to convince everyone that the world was going to end soon.
In those days the only conceivable way to destroy the world

would have been by flood. Great floods were common in the area and they would have provided a basis for the original story found in the myths of Babylon.

The Ark was seen as historically true until recent times.

In the 17th century, Athanasius Kircher set out to calculate how the Ark could hold all the Earth's land creatures. He translated the biblical 300 cubits into 135 meters, and drew a detailed plan to that measure, with three decks each holding 300 cubicles. Everythings fits perfectly - animals, food, waters, supplies, and Noah's extended family. Fortunately for him, Kircher only knew 340 kinds of animals. Today we know about 30 million. (Skehan, 1986, p.17

Kircher did not know of the existence of dinosaurs. Such knowledge would have demanded drastic changes in space requirements. Creationists in the twentieth century have not improved on Kircher's seventeenth-century Ark. If anything, theirs is less sophisticated.

Kircher's diagram shows how carefully he had studied the problem but modern science supports modern biblical scholarship in providing evidence that the story of Noah is most improbable. The biblical writers changed the mountains of Nizir to Mount Ararat. It is not possible to identify either area today.

The majority of people searching for Noah's Ark today are creationists. The Great Flood is of crucial importance to them in explaining the face of the earth. To find the remains of the Ark would be equivalent to King Arthur's Knights of the Round Table finding the Holy Grail.

The utter religious conviction that the Ark exists tends to blind creationists to biblical and scientific evidence. Such conviction – however commendable – leaves them open not only to self-deception but to the likelihood of falling victim to the hoaxes of others.

There are countless books and articles which show that, in the light of modern knowledge, the Ark and the Flood are improbable events. None of these ever will change, nor presumably ever have changed, the views of a convinced creationist. Minds, like parachutes, work only when they are open. The Flood and its implications are of crucial importance to creationism. For the most part, creationists seem to be psychologically incapable of conceiving that they could be wrong, even when faced with the clearest of evidence.

It is hard to imagine how anyone could close their eyes to the evidence presented by Michael Archer, professor of biology at the University of New South Wales:

It is clear that there is no scientific evidence for a worldwide flood at any stage in the Earth's history. It is also clear that Earth's sedimentary rocks formed over a vast period of time. The oldest are approximately 3 700 000 000 years old while the youngest are still forming. These sediments include many formations not deposited by water (such as the massively thick deposits of desert sands) interspersed with some that represent marine deposits and others that represent freshwater deposits. Further, within these sequences are many successive erosional unconformities indicating major breaks (lasting millions of years) in the process of sediment accumulation.

In any case, the appearance and disappearance of the additional amount of water (4 400 000 000 km^3) required to cover the Earth's mountains, which is over three times the amount (1 370 000 000 km^3) presently contained in all the Earth's oceans, would have imposed simply impossible constraints on the pre-Flood creatures of Earth and the inhabitants of the Ark during its journey (Soroka and Nelson, 1983). If that much extra water fell as rain, the pre-Flood Earth had to have had an atmospheric pressure about 840 times higher than it has now and an atmosphere which consisted of 99.9% water vapour (which would, incidentally, have been unbreathable). Further, from a thermodynamic point of view, because 2.26 million joules must be given up as heat for each kilogram of water condensed out of the atmosphere (Soroka and Nelson, 1983), that much water vapour condensing into rain would have raised the temperature of the Earth's atmosphere in excess of 3500°C during the time of the Flood. The consequences for the occupants of the Ark in what would have been a boiling ocean and unbreathable atmosphere bear thinking about. Even if the extra water welled up from within the Earth, the temperature of subsurface waters of this volume, because of their closer proximity to the hot mantle of the Earth, would have resulted again in oceans boiling away at temperatures of approximately 1600°C. Either way, Noah's geese would have been cooked.

Other absurdities inherent in the study of Biblical "arkeology" (Moore, 1983; Schadewald, 1982) include the problem of survival of diseases. If all of the disease-causing organisms known to be obligate parasites of particular hosts survived the flood, Noah's lot had to be raging pesthouses. The humans aboard, for example, had to have all of mankind's specific diseases, fatal and otherwise, including (among dozens of others) measles, poliomyelitis, typhus, four different kinds of malaria, gonorrhoea, syphilis, smallpox, leprosy, typhoid fever and pneumococcal pneumonia. If they could have survived these diversely horrid diseases and infections, the Ark's beseiged occupants would have developed immunities. The host-specific diseases would then have died out because of a lack of non-resistant hosts. Clearly, without resort to miracles (God or the Devil having created disease-causing organisms after the Ark grounded) which Creation "Scientists" claim not to have to rely on to support the Genesis account, the whole thing becomes rather silly. If, on the other hand, they allow a second miraculous Creation to account for present-day diseases, there can be no longer even a pretence of science about the hypothesis because miracles are outside the realm of science. The third alternative explanation, that present-day disease-causing organisms evolved within the last 4000 years from benign ancestors which were on the Ark, demands acceptance by Creation "Scientists" of more rapid and substantial evolutionary changes than even evolutionists would allow. (Archer, 1987b, pp. 132 - 3)

Anyone interested in this aspect of the Flood would be well advised to read the whole section by Archer (1987b, pp. 132 - 7). Archer then considers biogeographic patterns – the way animals and plants differ between continents and regions. These are powerful evidence for evolution.

The evolutionary model would predict that organisms evolving through time would produce descendants that resembled their regional ancestors. No divine Flood, mysterious process or other miracle is required to explain this biogeographic data. Hence it comes as no surprise to evolutionists to discover that fossil koalas are only known from Australia. Even Darwin realised that biogeography provided important support for what he called the "Law of Descent". As early as the 1830s, he realised that the fossil animals of Australia most resembled the living animals of the same continent. In the same way, he discovered that the extinct animals of South America resembled the living animals of that continent. This only seemed to make sense if the older animals were in fact the ancestors of the younger animals. To Darwin and most modern scientists, these data supported a model of evolutionary descent with modification. (Archer, 1987b, pp. 134 - 5)

These problems are insuperable for the creationist model and are never mentioned in their writings. Archer examines the problems confronting a platypus if the Flood were actual:

To consider just one example relevant to Australia, contemplate the lot of the platypuses that left what must have been a carefully constructed platypusary on the stranded Ark (for they could survive in nothing less than this as many zoologists have demonstrated). In the first place, since Creation "Scientists" have declared that at the time Mount Ararat was an active volcano (LaHaye and Morris, 1976), the platypus pair would have had to survive intense heat and a poisonous atmosphere. Then, because

there are no platypuses (fossil or living) anywhere in the world except Australia, the surviving pair must have made their way across the barren post-Flood wastelands (remember, every living substance had been destroyed) of Eurasia. Their journey would have been arduous. There was no food, for the organisms comprising their diet would have had no time to multiply and spread (platypuses require prodigious amounts of quite specific types of food every day). Neither was there fresh water for drinking or bathing (they constantly need to bathe in fresh water). This simple requirement would have been frustrated since every body of water stranded on the continents would have been filled with the foetid carcases of all kinds of dead animals as well as a host of other pollutants.
Assuming that they survived the 24 000 km journey, on arrival in Australia they faced starvation and dehydration in a land more thoroughly devastated than Hiroshima. How did they survive to reproduce? (Archer, 1987b, p. 135)

The Flood is a lynchpin of creation science. Without it the whole edifice collapses. Yet the evidence against it is incontestable. And still they believe. The moral is that if you have any notions of converting creationists, forget it. Concentrate, instead, on keeping others out of their clutches, particularly the children in the schools of Australia.

Chapter 4

Gish the Debater

In 1980 Duane Gish was secretary of the Institute for Creation Research (ICR) in San Diego. By 1988 he was vice president, second only to Henry Morris, the founder of the institute. Even Henry's son, John, has not risen to such heights.

It is worthwhile spending a little time analyzing Gish's debates not only because such an analysis provides a range of information on the hoaxes of creation science but also because Australian creation scientists copy his strategy, techniques and arguments. He is a frequent visitor to Australia and in March 1988 various fundamentalist groups arranged public debates and talks against scientists in capital cities across the country. He claims to have engaged in some 200 such debates around the world. On Friday, 18 March 1988, it was Sydney's turn.

Gish's debates are predictable but effective. Although he usually has a majority of creationists in the audience, who come to see another scientist demolished at the hands of the master, there are other, probably more important, reasons for Gish's phenomenal success rate. One reason is that he invariably chooses to debate along lines such as "Creation or Evolution, the Evidence." This immediately places both evolution and creationism at the same level in the eyes of the public. At the same time, it strengthens the facade that creation science is a valid scientific theory and has nothing to do with the Bible or evolution. Thomas Jukes describes the debating techniques of Gish:

> Second in command of ICR is Duane Gish, PhD, University of California, Berkeley, 1952. His activities as Morris's protagonist have been likened to Thomas H. Huxley's role as "Darwin's bulldog". It is over-imaginative to compare Morris to Darwin, but, nevertheless, Gish peripatetic and tireless, glib and unabashed, has elevated himself into prominence by incessantly reciting a rehearsed attack on evolution. In so doing, he poses as an authority in several fields, but says that "the Creator used processes that are not now operating anywhere in the natural Universe". When confronted with scientific objections to supernatural scenarios, Gish says that his religion is being attacked ... Gish asserts that "scientific creationism" is indeed science, but Judge Overton noted that Gish also said:
> "Creationists have repeatedly stated that neither creation nor evolution is a scientific theory (and each is equally religious)." (Jukes, 1984, p. 398)

On 9 April 1982, Gish spoke to a capacity audience at the University of California, Berkeley. The audience abounded with scientists and Gish was severely defeated He has never returned to Berkeley. On 18 March 1988, Gish debated Ian Plimer at the University of New South Wales. (Plimer is head of the department of geology at the University of Newcastle.) For this debate the audience was predominantly creationist and at times the hall was in an uproar. Gish was again severely defeated.

There are a number of parallels between the two events.

Uncorrected Credentials?

Thousands of one-page handouts were circulated by creationists who sponsored the Gish/Plimer debate. In this Gish is described as

a *Phi Beta Kappa* of the University of California, Los Angeles,

where he graduated with a Doctorate in Biochemistry. He is a Fellow of the American Institute of Chemistry, having had no less than 40 published papers in standard scientific journals.
Dr Gish spent 18 years in scientific studies at Cornell University Medical School, and was involved in research at the Upjohn Company where he worked under two Nobel prize winners.

This circular was on the letterhead of the National Alliance for Christian Leadership, dated 2 March 1988, with address in Canberra. It is signed by John Heininger, National Executive.

Gish's 1982 appearance in Berkeley was in the form of a lecture. Scientists would not debate on the grounds that to do so would appear in the eyes of the public to raise creationism to the level of a science. The audience overflowed the hall and comprised a majority of scientists from the staff at Berkeley. Pamphlets countering Gish were also distributed at the meeting.

The pamphlets featuring Gish's credentials at the Berkeley lecture were issued by the co-organizers of the lecture, Collegians for Christ and the Chinese Bible Church. The pamphlets at Berkeley in 1982 gave the same credentials for Gish as were given in Sydney in 1988. This gave rise to charges of falsification.

In addition, Berkeley medical physics Professor Thomas Jukes has charged creationists with falsifying Gish's credentials. A flyer announcing the lecture claims Gish "spent 18 years as a faculty member at Cornell Medical School, and is ... a member of Phi Beta Kappa.
Jukes asserts that information from the biography of "American Men and Women in Science" makes it "impossible that he spent 18 years at Cornell." Gish spent only two years at Cornell as a postdoctoral fellow instead of a professor, Jukes said, and there is no record of Gish as a Phi Beta Kappa member. (Ciraolo and Knobbe, 1982, p. 1)

In the course of an interview with a reporter after the lecture the question of qualifications was raised.

Flyers advertising the UC Berkeley lecture stated that Gish spent 18 years as a faculty member at Cornell University, but he said this was "an error." He spent 1953 through 1956 as a postdoctoral fellow at Cornell, and was an assistant professor there in 1956.
Since 1956 he has worked as a researcher for the Upjohn Company, as a professor at Christian Heritage College, and as associate director for the Institute for Creation Research. (Ciraolo and Knobbe, 1982, p. 1)
It would seem difficult for Gish to deny that since 1982 he has knowingly allowed the circulation of inaccurate credentials. One can only wonder at how this started in the first place and how long it will continue.

Brainwashed?

In 1988 Plimer pointed out five serious errors in fifty-five words, and eighty-seven errors all told, in Gish's comic book *Have You Been Brainwashed?* – "a lie every 11 words!" Gish replied that it was written a long time ago, before it was known that fossils existed in the earth's crust and Precambrian rocks. Plimer asked why the booklet was still being sold "at this very moment" in the foyer. Gish had no reply.

In 1982 similar charges were made against Gish at the Berkeley lecture. *Gish alleged then that "someone else wrote the booklet"* (Jukes, 1984, p. 399). The booklet was first published in 1974, and as long ago as 1982 Gish claimed that this and his two other creationist books *Dinosaurs: Those Terrible Lizards* and *Evolution? The Fossils Say No* "have a distribution of over two million copies" (Jukes, 1982). There has not been one word changed in the *Brainwashed* booklet, nor have any of the errors acknowledged by Gish years ago ever been removed. Gish's name is on the front cover, and on the back cover of early editions readers are urged to buy "a copy of the complete book by Dr. Gish entitled *EVOLUTION: The Fossils Say No!* " The back covers of later editions do not state its origin as from Gish's evolution book. They do say: "The wonderful news in the booklet you hold is just too good to hold to yourself."

Gish would be hard put to deny charges that he is deliberately publishing falsehood in this booklet and in the book from which it is derived. There are hundreds of thousands of dollars in royalties involved and the books are still making money hand over fist.

When Is Religion Religion?

In 1988 a question was asked about why the major Christian faiths and all biblical scholars and theologians opposed creation science's interpretations of the Bible. Gish forgot his public image that creation science was supposed to be science and independent of religion. He leaned toward the front part of the hall, crowded with creationists, and asked "Why are the big churches always attacking us little churches?" He avoided answering the question.

Jukes describes what happened in 1982 when Gish was asked a question by Steve Obrebski, Ph.D., now a scientific consultant at San Francisco State University:

"The director of your institute (Henry Morris) has claimed that there are concentrations of good and bad angels around the Earth, that UFOs are manifestations of the Devil or the dark forces, that our inanimate souls can travel faster than light, that although stars can't affect us, the evil spirits associated with them can, and that the scars on Mars and the Moon, the asteroids and the rings of Saturn and meteor showers are a reflection of some big turf battle between Satan and Michael and angels. And you have said yourself that there was no problem of incest among the progeny of Adam and Eve because God made them genetically perfect. Now I ask you sir ... what are the facts? How would you disprove them? Where do you get the material evidence for those facts, and how can you claim to be doing science when you make such wild and tentative assertions and then come in here and mix both sciences and mix up what the evidence is and claim that you are doing creationist science?"
This was followed by prolonged applause in the form of a standing ovation.

Gish's response was, first, to make a diversion, alleging that someone else wrote the booklet, "Have You Been Brainwashed", published under his authorship, and then to say:

"I have not gone into this matter of religion. I have stuck strictly to science here tonight. It is always the evolutionists who want to bring in the matter of religion. This man has been ridiculing the Christian faith."

The connection between the Christian faith and the fantasies of Henry Morris is dubious, and, of course, it is the creationists who bring in the matter of religion: Gish in his booklet *(Brainwashed)* threatens his recalcitrant readers with terrible judgment and says, "I urge you to make your peace with God today". (Jukes, 1984, p. 399)

Creation Science Can Be Public or Christian But Not Both

In 1988 Plimer challenged Gish on the illegal use of public funds to teach creation science to children at Smith School in Livermore, California. The eight weeks' course consisted of brainwashing techniques for fifth and sixth grade classes, including six gifted children for whom the funds were originally intended. Implicit throughout the course, which was based entirely on creation science texts, films and guest speakers, was that a choice must be made between creationism and atheism.

Plimer read from original documents which included student worksheets and tests. He showed the audience excerpts from Gish's books which the children had been forced to read.

At the end of the course the class was forced to vote *either* CREATIONISM *or* ATHEISM. The six gifted children voted *ATHEISM*. The teacher sent them back to the library to go over the films again!

Plimer challenged Gish that his materials turned out atheists. Gish was highly embarrassed over this but claimed that the teacher had acted without his permission and, in addition, had used materials "intended for Christian schools not public schools." An examination of the lists of the materials and borrowing cards indicates that this is simply untrue. The most popular texts used were those by Gish, *Dinosaurs: Those Terrible Lizards*, and his colleague Gary Parker, *Dry Bones ... and Other Fossils*. The materials for public schools to which Gish refers are simply standard creationist books with biblical quotes removed so that they have the appearance of science books not based on the Bible.

The above amply illustrates Gish's proven ability not merely to twist but wrench apart the truth when cornered. Besides that, does Gish believe that all students at Christian schools are fools? And does he believe that some of them, in spite of persistent brainwashing, will not see through the fraud and sooner or later choose atheism? What he is admitting is that what is deemed to be fraud in public schools is all right in Christian schools. Surely there are at least some parents of children in Christian schools who would object to this if it became known. A more detailed treatment of the Livermore episode is given in Chapter 12.

The Almighty Dollar Rears Its Ugly Head

When Gish was accused by Plimer of "being in it for the money" he protested loud and long the financial sacrifices he had made in leaving his employment at Upjohn, including the massive loss in retirement benefits. Like a monk sworn to a vow of poverty he described how every cent he received in appearance fees went to the Institute for Creation Research whereas Plimer was free to pocket his fees. What he did not explain was that the ICR is essentially a marketing organization and that the proceeds from the sale of his books provide an annual income at least several times that of Plimer. In addition there are the perks attached to being vice president of an institute that is very solvent indeed.

Let There Be Light!

In recent years Barry Setterfield, an Australian creationist, has developed a theory on the slowing down of light which has become one of the cornerstones of the creationist "proofs" that the universe is only 6000 years old. This theory denies that light has taken millions of years to reach us (see Chapter 9). Setterfield's theory has become beloved of American creationists and much quoted in both Australian and American creationist literature. It has never been accepted for publication in any reputable scientific journal.

When Plimer demonstrated, in a way that even the creationist audience could understand, that Setterfield's theory was utter nonsense, Gish had no hesitation in saying that he never believed it anyway. Thus Setterfield was unceremoniously thrown to the dogs with no more hesitation than the Livermore teacher had been discarded a few years previously. But there is as much probability that Setterfield's erroneous book will be withdrawn from publication as there is for the withdrawal and acknowledgment of errors in Gish's *Brainwashed* booklet. The probability is zero for both, as it is for the dozens of other creationist books with known errors.

It is not for nothing that it has become a byword that if all the errors and misquotes were removed from creationist writings, there would be nothing left.

Two World Views?

Most of the audience at the Gish/Plimer debate missed the clever sophistry in Gish's opening remarks. He commenced by saying that "evolution and creationism are two world views."

This is false. The two world views are that of a dynamic, changing, evolving universe and that of a static, unchanging universe. Creationism is one religious interpretation of one particular world view that was current in the Middle East when the Bible was written. The cosmology in which Genesis is framed preceded and is independent of Genesis. Dozens of other religious interpretations of a static world view are present in creation stories in the religions of the world. Gish has no valid reason for preferring one creation story over another except that of prejudice toward the Christian religion. Like all creation stories it is a vehicle of religious truth using the cosmology and world view of a particular culture. It is not a scientific account of how the world was made.

Who Laid the Egg?

A further sophistry in Gish's opening remarks for the 1988 debate was to ridicule and discard almost all evidence from astronomy, the life cycles of stars, the decay series of the elements, the age of the universe, the age of the earth. Any one of these is sufficient to refute a 6000-year-old universe. Instead Gish chose to state that science believed that the universe originated in a cosmic egg which hatched primeval hydrogen that somehow formed the human brain. What's more, he got away with it! In so doing he attempted to deflect the debate into problems concerning the origin of life – an issue quite separate from the evolution of the universe.

Statistics Are What You Make of Them

After carefully lulling the audience away from the most cogent evidence for evolution in every other science, Gish confined the debate to the limited area of biological evolution and the origin of life. He then proceeded to brandish his favorite weapon, one that is much copied by his fellow creationists. He proved statistically that evolution is impossible!

In doing this he cited the life cycle of the monarch butterfly, a common example of metamorphosis in which an animal goes through several very different forms during its life cycle. The butterfly proceeds from egg to caterpillar to a cocoon filled with a jellylike substance. From the cocoon hatches the mature butterfly that has a life span of a few days within which to mate and lay eggs, whereby the cycle is continued. This is surely one of the many wonders of the living world.

Gish states that the number of simultaneous mutations required for an animal such as the monarch butterfly to evolve is so improbable as to verge on the impossible. Moreover, there are many such examples in nature.

Gish claimed that the only reasonable explanation is the direct intervention of God. Invoking God as an explanation for an area of science not yet understood is known as the "god of the gaps" theory. There are a great many processes which seemed beyond human comprehension at the time but which later yielded to sustained research. A well-known example from physics is that for decades scientists had to live with the apparently irreconcilable fact that light was both a particle and a wave.

However, the mechanisms of evolution may well be debated forever. For many, moreover, the existence of apparent mysteries like the life cycle of the monarch butterfly merely illustrates the inadequacy of neo-Darwinism as a full explanation of the mechanism of evolution. But little by little more light is being shed on these apparent mysteries. In butterflies

> the changes from a larva to a pupa are under control of two quite simple chemical substances. One of them is the juvenile insect hormone, and the other is ecdysone. Ecdysone changes the larva into a pupa. It is a hydroxylated cholesterol-type molecule. What happens, of course, is that some genes are turned on and others are turned off. All of the information for the butterfly is contained in DNA, and, since it is in DNA, it can have been produced only by evolution. That is, unless you believe that during the week of creation, several million DNA molecules, each containing up to 3 billion nucleotides, were arranged in specific sequences. (Jukes, 1988)

What Gish is also implying is that his little god cannot create through evolution because it is too complicated for him. The obvious rejoinder is that he should seek another god. In his case he could try the God of the majority of Christians, the one who created through evolution.

On the other hand, when Gish states the inadequacy of a biological viewpoint which is totally reductionist, he is doing no more than parroting the view of a great many scientists. Most scientists today would be dubious of total reductionism but unwilling to pull miracles out of a hat, as does Gish.

Gish's Thermodynamics

In his 1988 debate with Plimer, Gish trotted out another of his aces. This is his version of thermodynamics which is accepted as gospel by creationists and enshrined in the creationist film "Origins.

Gish gave the standard introduction to thermodynamics which teachers use on their students. He said that spontaneous change leads to greater disorder, that we don't see a pile of bricks suddenly of their own accord become a house. Rather, all nature tends to decay, become more disordered and die.

Gish then gave the usual argument of creationists: that evolution is impossible according to the second law of thermodynamics. Evolution states that the universe began simply, as hydrogen and energy, which then became more complex forming galaxies, solar systems and eventually life. Thus evolution states that the universe proceeded from simple to complex. This contradicts the second law; therefore evolution never occurred. Mind you, he *sounded* very convincing.

His argument is rather like expounding Newton's laws of motion but deliberately omitting one or two which may not be palatable. It sounds good but that's about all. It is about as useful as that three-legged stool with one leg missing.

What Gish omitted was that the second law holds only under special conditions. The second law does not hold, for example, while the sun still has a supply of hydrogen to feed energy into the planet earth. The sun provides the energy to keep, as it were, the second law at bay and so enable life to exist. Since Gish has misrepresented the second law in some 200 debates and many books it is difficult indeed to accept that he is not doing it deliberately. For a fuller discussion of this, see chapter 10.

When Your Friends Let You Down

If the reader thinks this treatment has been unnecessarily harsh on Gish, consider what fellow creationist John Robbins says about creation science depositions to the U.S. Supreme Court in 1987. Gish was a leading figure in preparing the creationist case which sought leave to have creation science admitted to schools as science, using the strategy of omitting all references to God in their writings.

Tom McIver gives the following summary of remarks by creationist John Robbins, head of the Trinity Foundation in Maryland, in his address "The Hoax of Scientific Creation" delivered at the 1987 Baltimore Creation Fellowship Conference:

He (Robbins) agrees with anti-creationists that so-called "creation-science" is a fraud and a deception. To pretend that creationism consists of scientific evidence and not religious concepts is a shallow, devious tactic doomed to failure. Robbins is appalled that Wendell Bird, the creationist lawyer, in an attempt to pass off creationism as merely science, declared to the Supreme Court that creation-science need not contain any concept of God or the Book of Genesis. Not only are they trying unsuccessfully to con the judges, they are conning Christians into supporting a movement that is betraying its very principles. "It is past time," says Robbins, "for Biblical Christians to consider whether they ought to continue to spend thousands of dollars on such specious arguments, and, more importantly, whether Christians can any longer afford to use a method of defending the faith that inexorably leads to non-Christian conclusions." (McIver, 1988, p. 11)

Gish is also chairman of the science and technology section of a creationist group known as the Coalition on Revival. Its manifesto demands acceptance of a Christian world view. The committee led by Gish supports increased military spending and proclaims that all of science must be based on the Bible (McIver, 1988, p. 14). Gish's position in this group seems difficult to reconcile with his proclamations around the world that creation science does not depend on the Bible.

However, Gish's presence on a committee which supports increased military spending is not inconsistent with the creationist belief that the Second Coming of Christ will be preceded by a nuclear war in which Israel defeats Russia.

How Not to Debate With Creationists

An analysis of the tapes of the Gish/Plimer debate and comparison with Gish's mauling at Berkeley in 1982 yielded interesting results, particularly in the areas of strategy in combating creationists as well as hints on what to do and what not to do when debating with them.

At Berkeley Gish refused to debate Fred Edwards, who had successfully debated Gish a few months earlier. Because of this no one else would debate with him and he was forced to switch to a lecture format, which resulted in his saying afterwards:

"The situation at Berkeley last night was totally unexpected," Gish said, as he settled into a chair in the corner of the lecture hall to discuss his life's crusade. "The behavior was the worst I've ever encountered. The audience showed a complete lack of consideration, and no desire to learn anything." (Knobbe, 1982, p. 4)

Gish has not debated or lectured at Berkeley since 1982.

After Plimer's 50-minute opening, Gish told the mainly creationist audience that in all his 200 debates around the world he had never been so disgustingly treated. Consequent upon this Carl Wieland, head of the Creation Science Foundation, wrote to Plimer telling him that as "a known troublemaker" he is banned from attending creationist seminars. It remains to be seen whether Gish will come back to Sydney.

Gish had a present waiting for him when he got home. Two television cameras hired by Plimer had recorded the whole debate. The tapes were rushed to America to an eagerly waiting audience, to be distributed for use on local TV and seminars. This is done through the Committees of Correspondence which are networked across thirty-seven states at the grass-roots level. The Committees of Correspondence are sponsored by the National Center for Science Education and affiliated with the American Association for the Advancement of Science.

Worthy of comment are Gish's opening remarks and strategy which has made him the champion debater of all creationism and earned him the sobriquet of "Morris's bulldog."

His very first words claim that "evolution and creationism are two world views." People in the audience nod in agreement. In fact creationism is one religious interpretation of a static, unchanging small cosmos which is earth centered. Evolution is a world view of a dynamic, changing, unfolding universe of immense size. Gish has immediately lulled his audience into overlooking the vast discrepancy between the two views.

In the next sentence Gish states that the debate centers around scientific evidence or rather "which is more likely, evolution or creationism." He states that the debate is not about the Bible, theology, the age of the earth or the distance of the stars, but about *evidence*. In so doing he subtly disengages from all the areas where there is the most convincing evidence for evolution. All these are put out of bounds – the age of the universe and the solar system, the evidence from astronomy on the birth and death of stars and the vast distances, the nigh incontrovertible evidence from optical astronomy that galaxies are 2 million light-years distant, the decay series of the elements whereby all known elements are formed from primeval hydrogen. The literal interpretation of the Bible as science is put out of bounds. Gish defines the debate to a narrow area which he can control.

If the unsuspecting opponent, overconfident perhaps in his mastery of a narrow area of science, accepts Gish's boundary conditions, then he is already doomed. I say "he" because there are no women debaters against creationism. The women scientists don't seem to debate against creationists for whatever reason. Perhaps they have more sense?

Once the boundary lines are set, Gish, or any of the other competent creationist debaters, can give fifteen or more well-rehearsed arguments making nonsense of evolution. This can be done with conviction, skill and humor in such a way as to not only get the audience on side but to subtly implant the view that creationism is the way used by God and that evolution is used to disprove God.

In reply the luckless scientist can, in the time available, refute at best only two or three of the creationist arguments. The majority remain unanswered in the minds of the audience. The field of debate being limited, the topics which remain are usually the fossil record, transition species, the origin of life and such other areas where the evidence may be very powerful indeed to a scientist but difficult to convey to a lay audience in the confines of a debate. This is especially true when the creationist opponent is prepared to misquote and is well-armed with rehearsed answers to most of the points likely to be raised. The champion of truth is amazed to find that he has lost on points and is now notch 201 on Gish's belt.

What Plimer did was to ignore the rules, to risk the frowns of the academic community, to attack Gish on every known angle, including his record of proven lies. Not very nice but very, very effective, especially when videotaped for an international audience.

The alternative is to refuse to debate, on the grounds that the very act of consent elevates creationism to the level of science. Creationists wish to debate scientists at every possible opportunity. Surely it must be obvious that they would not do this unless the past record showed that the odds were very much on their side. To force them to lecture rather than debate to get the message across is an alternative which evens the odds. Undoubtedly this is the best tactic unless there is someone who understands the strategy and who is prepared to do the homework across a variety of areas and also prepared to attack head-on.

There are exceptions but in the main these are the alternatives: debate and probably lose, or refuse to debate and force a lecture. By a thorough analysis of previous debates and tactics, guidelines can be drawn up for debating with creationists in Australia, as has already been done in America. The bibliography lists a few resources for this effort.

Chapter 5

New Worlds For Old

There is a view in which the universe begins with an enormous explosion of light and heat. This moment was the beginning of time and space. Over unimaginable eons of time the outrushing energy froze into matter and gave birth to stars and galaxies. Ten billion years or so later, in a remote corner of a remote galaxy, the jellied energy we call matter showed an even more remarkable potential. This was the beginning of life. Thereafter the pace speeded up. Life became more complex by developing the ability to sense and respond to its surroundings.

The complexity of forms multiplied and some were discarded while a few achieved consciousness. There was now a web of life-forms intricately linked and dependent on each other as well as on the planet itself. Almost yesterday came the final denouement. A life-form conscious of being conscious appeared. Having reached its climax, purpose and end point, this unique speck in the universe was handed over to the care of this being with self-consciousness, the being that we call man. One of the qualities which sets man apart is his overwhelming urge to seek an answer to his origin, the search for meaning.

Poetry, mysticism, religion? Perhaps. But the big bang theory, as it has been named, is now taken very much for granted by astrophysicists (Davies, 1983, p. 22). The echoes of the big bang were detected experimentally in the mid-1960s.

Space-time, antimatter, quarks and all the other strange particles and laws of modern physics should have prepared us for a creation story stranger than anything the mind of man could possibly have imagined.

In addition, the more we find out about the planet and the many stringent conditions necessary to maintain it as a "life support system," the more improbable seems life. Nuclear destruction may be a possibility but the end of life on earth as we know it seems certain if pollution and laying waste to the environment continue at the present rate. So fragile is life.

The cosmos has evolved from the simple to the complex, from small to large. It has evolved irreversibly. Some 3.5 billion years ago a new phase of evolution occurred on this planet earth. Inanimate matter revealed the potential within itself to become life. It was a potential not realized anywhere else in this vast universe, to the best of our present knowledge.

The great physicist and Nobelist Max Born stated in 1951:

The scientist's urge to investigate, like the faith of the devout or the inspiration of the artist, is an expression of mankind's longing for something fixed, something at rest in the universal whirl: God, Beauty, Truth.
Truth is what the scientist aims at. He finds nothing at rest, nothing enduring, in the universe. Not everything is knowable, still less predictable. But the mind of man is capable of grasping and understanding at least a part of Creation; amid the flight of phenomena stands the immutable pole of law.

> 'Tis thus at the roaring Loom of Time I ply,
> And weave for God the Garment thou seest Him by.
>
> Carlyle's translation of Goethe
> (Born, 1969, p. 166)

Creationism a World View?

Creationist writers and speakers consistently tell us that we must make a choice between two world views. On the one side, they say, is evolution, which has no need of a god, and on the other side is creationism, which says that God directly created the world 6000 years ago. But is creationism a world view?

The world view of the ancients is simply the way the world appears to our senses. It is sufficient for survival and very much what we use in most of our waking lives. We walk or drive between towns or build a home without giving a thought to what we learned in childhood – that the earth is round not flat. When Magellan circumnavigated the globe his sailors were afraid of sailing over the edge of the earth. We still speak of the four corners of the world. In the early seventeenth century, fundamentalists opposed a round earth as being contrary to Scripture. Muslim fundamentalists and a number of creationists still do. Because our senses and experiences of nature tell us that all solid objects must rest on something which is also solid, ancient philosophers put forward many different concepts of what supported a flat earth. In 6 B.C. the Greek philosopher Thales imagined the earth as a flat disk floating on water, the basic stuff of the universe from which all else emerged. The ancient Hindus imagined the flat earth supported by four elephants standing on a turtle swimming in a sea of milk.

The ways in which we picture the world and cosmos contribute to the world view, in any culture in any age. It is apparent that a world view may have religious elements inserted into it and in nearly all cultures past and present this has been the case, as in the Hindu world view. But it is not totally dependent on beliefs of how the world came into being. To say that a god or gods created the world is quite a separate issue. It is a theory of the origin of the world and is not essential to the world view.

Scarcely 200 years after a round earth became accepted Copernicus proposed a changed world view which goes even more strongly against our intuition. His proposal that the earth is not the center of the universe is against the evidence of our senses. We *know* that the sun goes around the earth. We still speak of "sunrise" and "sunset." More than this, it shifts earth and man away from being at the center of the universe and it makes man not the most important being in creation.

This new way of thinking, this new world view, languished as an interesting speculation until the time of Galileo and Kepler but it required the awesome genius of Newton to put it all together. The new world view of Newtonian physics covered not merely planet earth but reached out to the far boundaries of the universe. It was the experimentally verified predictions of Newton which made it impossible to deny this new way of thinking about the world. It was a world view which now encompassed the planets and the stars and very soon thereafter the distant galaxies.

But Newton effected much more than a new world view. Aristotelian physics had taught that everything in nature had its place, that heavy objects fall faster than light objects – which is what our senses tell us most of the time. Newton revealed that every particle of matter in the universe attracted equally every other particle, that matter was "democratic." While Aristotle's theories of physics justified an aristocracy, Newton's did not. The old physics had been used to argue that God made some men to be rulers and others to be ruled. Revolt against the King was revolt against God. Hence the "divine right of Kings."

The French philosophers were quick to see that Newton's physics demolished the King's divine right. The physics of Newton was one of the major inputs into the ideas and events which led to the French Revolution. The same concept was enshrined into the American Declaration of Independence: "that all Men are created equal, that they are endowed by their Creator with certain inalienable Rights, that among these are Life, Liberty, and the Pursuit of Happiness."

Basil Willey lists just a few of the ideas consequent upon the Newtonian synthesis, or the "New Philosophy" as it was called:

First, it produced a distrust of all tradition, a determination to accept nothing as true merely on authority, but only after experiment and verification. You find Bacon rejecting the philosophy of the medieval Schoolmen, Browne writing a long exposure of popular errors and superstitions (such as the belief that a toad had a jewel in its head, or that an elephant had no joints in its legs), Descartes resolving to doubt everything – even his own senses – until he can come upon something clear and certain, which he finally finds in the fact of his own existence as a thinking being. Thus the chief intellectual task of the seventeenth century became the winnowing of truth from error, fact from fiction or fable. Gradually a sense of confidence, and even exhilaration, set in; the universe seemed no longer mysterious or frightening; everything in it was explicable and comprehensible. Comets and eclipses were no longer dreaded as portents of disaster; witchcraft was dismissed as an old wives' tale. This new feeling of security is expressed in Pope's epitaph on Newton:

> Nature and Nature's laws lay hid in night;
> God said, *Let Newton be!* and all was light!
> (Willey, 1959, p. 109)

The new world view developed from Newtonian physics and the "New Philosophy" no longer had earth at the center of the universe. The universe was vast but predictable. Heaven – the abode of God – was no longer a geographical place just beyond the visible firmament. But it was still the world view of a static universe. It was a clockwork universe with God the winder of the clock. There seemed no place for free will. But this was not to last very long.

The final attack on our sense-derived intuition of the world about us was yet to come. A flurry of scientific activity preceding the twentieth century gave us the world of atoms which by early in the twentieth century had become miniature solar systems comprising mostly empty space. Remove the space, and you and I could sit together comfortably on the head of a pin. These atoms vibrate constantly while they touch each other and even within themselves. They are never still. At the same time, both physics and geology were showing that the age of the earth was measured in millions of years, not thousands. After years of hesitation Darwin finally published his book. Alfred Russel Wallace, another British naturalist, had arrived at the same conclusion independently. The origin of existing life-forms through evolution became accepted.

The twentieth century has seen a deluge of new discoveries confirming that we are part of an ever changing, ever evolving universe. Space and time are now parceled together, although on reflection we know that we have never seen an object existing in space that did not exist in time and by the same token we have never known time to exist by itself, that is, without some change occurring in objects whether they be changes in metabolism within our own bodies or in things in the world around us. The exception is the timeless time which is in our dreams and is located in the right hemisphere of our brain. It is this timeless time which is a part of the Aboriginal Dreamtime stories and also the basis of many creation stories around the world. When we commence a story with "once upon a time" we move our listeners away from the everyday time ticking away on the clock.

The constancy of the speed of light appears to defy all common sense. The evolution of life is but part of a much wider evolution of the whole universe. The echoes of the initial "big bang" have been measured and particles of matter have been created out of energy in the laboratory.

Above all, the clockwork mechanical universe of Newton has gone forever. The flat-earth theories are only the dimmest memories of our early childhood. In prying open the secrets of the atom and the codes of life we have opened a Pandora's box which will either destroy us or give to humankind a future as yet undreamed of. While our children's children may one day view us as old fossils who had difficulty comprehending the obvious it may be of some consolation to know that our difficulties are shared by many scientists themselves.

The pathologist, writer and television producer Kit Pedlar describes his astonishment, while planning a television series on the mind and the paranormal, at coming across the concern that modern physicists have for broader issues:
"For almost twenty years I occupied my research time as a happy biological reductionist believing that my painstaking research would eventually reveal ultimate truths. Then I began to read the new physics. The experience was shattering.
"As a biologist I had imagined the physicists to be cool, clear, unemotional men and women who looked down on

nature from a clinical, detached viewpoint – people who reduced a sunset to wavelengths and frequencies, and observers who shredded the complex of the universe into rigid and formal elements.

"My error was enormous. I began to study the works of people with legendary names: Einstein, Bohr, Schrodinger and Dirac. I found that here were not clinical and detached men, but poetic and religious ones who imagined such unfamiliar immensities as to make what I have referred to as the 'paranormal' almost pedestrian by comparison." It is ironical that physics, which has led the way for all other sciences, is now moving towards a more accommodating view of mind, while the life sciences, following the path of last century's physics, are trying to abolish mind altogether. The psychologist Harold Morowitz has remarked on this curious reversal:

"What has happened is that biologists, who once postulated a privileged role for the human mind in nature's hierarchy, have been moving relentlessly toward the hard-core materialism that characterized nineteenth-century physics. At the same time, physicists, faced with compelling experimental evidence, have been moving away from strictly mechanical models of the universe to a view that sees the mind as playing an integral role in all physical events. It is as if the two disciplines were on fast-moving trains, going in opposite directions and not noticing what is happening across the tracks.

In the coming chapters we shall see how the new physics has given "the observer" a central role in the nature of physical reality. A growing number of people believe that recent advances in fundamental science are more likely to reveal the deeper meaning of existence than (that which would) appeal to traditional religion. In any case, religion cannot afford to ignore these advances. (Davies, 1983, pp7-8)

It is apparent that the creationist claim that evolution and creationism are just two world views is at best dubious and at worst sophistry.

Creationism stands in contrast to theistic evolution. Creationism claims that God created the world as described in the Bible; theistic evolution claims that God created the world through evolution.

Evolution is the central unifying concept of a world view which sees the universe as dynamic and evolving. In contrast to this view are many world views of ancient times which see the world as static and unchanging. Creationism has appropriated one of these ancient world views, namely the world view current in the Middle East in about 1000 B.C.

Creationism is not therefore a world view. It is a statement that God made the world according to a world view now largely discarded.

Similarly, evolution is not a world view by itself, although sometimes used that way. Evolution is a concept which unites all branches of science. Evolution in its widest sense includes:
• stellar evolution (astronomy),
• the formation of elements from primal hydrogen (chemistry),
• the way in which life developed (biology),
• the formation and changes which have shaped the surface of the earth (geology), and
• the understanding of the processes by which mass-energy initiated the universe (physics).

These are, of course, the traditional divisions of science. There are nowadays hundreds of different branches and sub-branches in science and much overlapping but the above indicates why evolution is considered central to all the sciences although more widely used in some than others.

The world view today is that of a dynamic, evolving universe of immense age and unimaginable size. It has been described as a restless universe, where stillness is a word for something which is not present in any part of the universe. It does not seem possible to take us any further away from the intuitive world, based on our senses, in which we live and breathe and have our being. Perhaps there is always a tinge of regret at leaving behind the simple black and white world of our childhood.

The World View in Biblical Times

The creation science view of the world is based on the biblical perception of creation.

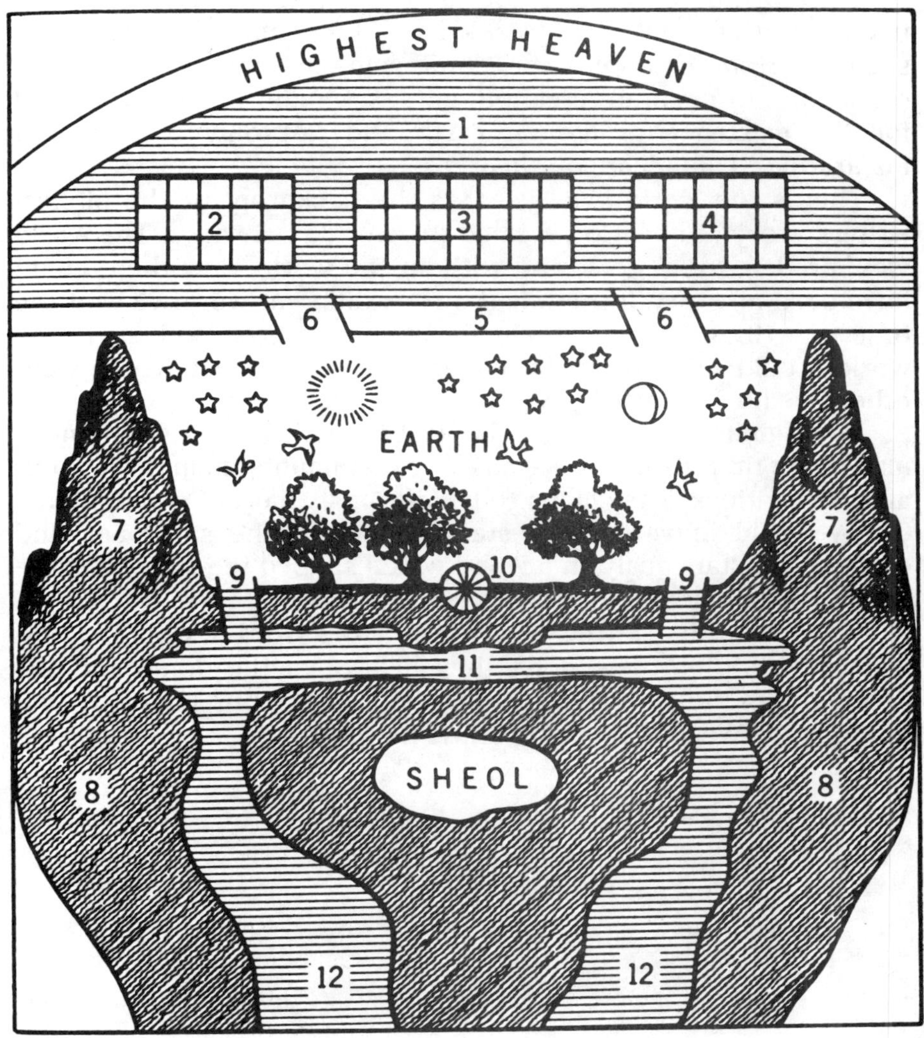

Biblical conception of the world: (1) waters above the firmament; (2) storehouses for snows; (3) storehouses for hail; (4) chambers of winds; (5) firmament; (6) sluice; (7) pillars of the sky; (8) pillars of the earth; (9) fountain of the deep; (10) navel of the earth; (11) waters under the earth; (12) rivers of the nether world. (From Frye, 1983, p. 159; reproduced with the permission of Charles Scribner's Sons.

This world view is static and changeless. The earth is unchanging and flat. The world exists for mankind, and man is separate from creation. The stars, sun and moon are merely ornaments to decorate creation, not, as we know today, essential for the existence of life on earth and not only a source of inspiration to illuminate our hearts.

This is the world view used as the backdrop for the biblical writers' story of how God created the world in six days. The world view predated the story, and without the world view there could have been no story. If there had been a different world view at the time of writing Genesis, then the theory of how God made the world would necessarily have been written in that world view, otherwise no one would have understood it.

It is significant that it is in biblical cosmology that creationists make their first major error – an error that haunts them throughout all their writings. They claim to believe in six days of creation, in a 6000-year-old universe with stars and earth the same age and created simultaneously, in a Flood which shaped the surface of the earth. And so on.

They claim this is true "because it is the Word of God and God cannot lie." Why, then, don't they believe in the dome and the floodgates of the sky, pillars of the earth, and so on? These are just as much the Word of God as six days of creation.

Chapter 6

The Bible as History

The biblical timeline of about 2000 years from Abraham to Christ is generally agreed as historically accurate, although the exact dating of events and the way in which they actually occurred are hotly debated among archeologists and biblical scholars. The times before Abraham are labeled as "prehistory" in modern Bibles.

Scholars agree that Genesis, although the first book of the Bible, was one of the last books to be written (Price, 1986a, p. 2). It reached its final form around 450 B.C. and one purpose was to complete the story of the Israelites, starting with historical Abraham and working back to the beginning of the world. Another purpose was to show that the God of Israel was also the God of all nations and of creation itself (Clancy, 1968).

Even in the historical parts of the Bible which commence with Abraham, the Jews were simply not interested in writing history as we know it. The authors did not merely report events, they changed details and highlighted certain aspects of the events. This was done to illustrate their total conviction that the God of Israel *intervened in history* on behalf of His Chosen People.

The "prehistory" in the first eleven chapters of Genesis is a collection of ancient stories and myths with which the audience was familiar; some were from neighboring civilizations and some from unknown sources. The final editors did not hesitate to put conflicting stories side by side. However, the biblical authors drastically changed the meanings of the original sources while making them serve as a vehicle for what they perceived as religious truths concerning the God of Israel.

Genesis 1 to 11:26 describes the events from the first day of creation to the birth of Abraham. Chapters 5, 10 and 11 give genealogical lists which enable the calculation of the years from creation to the birth of Abraham, the father of the Jews. Such lists of names were a common literary device used to emphasize the ancestry of a particular person. In this case they showed that all the neighboring races had the same ancestry. In other words, all human beings were created by the God of Israel. This was in direct contravention of the beliefs of neighboring civilizations who claimed to be created by their own particular god or gods. In the culture and at the time of writings, such lists would not have been taken literally but everyone would have understood the point being made.

Ussher's Chronology

Anglican Archbishop James Ussher (1581 - 1656), a biblical scholar of great repute, is best remembered as the person who used the Bible to calculate the date of creation as 4004 B.C. His chronology of the Bible was accepted and followed for many years.

Although creation science generally follows Ussher's chronology, with a few notable exceptions which are discussed in later chapters, there is divergence among creation science researchers concerning the exact dates. For example, the date of creation varies between 4004 B.C. and 4000 B.C. The date of the Flood is given as 2345 B.C. by Setterfield (1986, p. 172) and as 2304 B.C. by Osgood (1984, p. 141). But, overall, the discrepancies are usually no more than a few years.

Creation Science Chronology of Genesis

4000 B.C.: The Six Days of Creation
- The first pair of humans came into existence as adults. They lived in the Garden of Eden. They were immortal.
- The speed of light was 200 billion times faster than it is today.
- The second law of thermodynamics did not exist.
- There was no pain, death or suffering for any living creature.
- All living and extinct species of life were in existence. This included the many species of dinosaurs.

3870 B.C.: The Birth of Seth, a Son of Adam
- Seth was born when Adam was 130 years old. This was after Cain killed Abel.
- Adam and Eve had disobeyed God and were driven from the Garden of Eden prior to the birth of Cain and Abel. Hence the upper limit of Adam and Eve's habitation in the Garden may have been 3900 B.C.
- Pain, death, suffering and the second law of thermodynamics were now in existence.

2345 B.C.: The Flood
- All life on earth was destroyed when a great flood covered even the highest mountains.
- All landforms and oceans on the planet today were formed by the Flood.
- All fossils are a result of the Flood.
- Noah and his family survived the Flood by building the Ark, a large wooden vessel, and living on it until the water receded, about one year.
- Noah took on board at least a pair of every species of plant and animal. The animals included hundreds of thousands of insects, as well as all the reptiles and dinosaurs, some up to 30 meters (about 98 feet) in length. There were many natural predators and meat eaters on board.
- *All life-forms on earth today are descended from those on board the Ark .*

2195 B.C.: The Tower of Babel
- In the 150 years since the Flood, Noah's family had grown to a very large number and they all spoke the same language.

(continued)

(continuation of Creation Science Chronology of Genesis)

• They co-operated in an attempt to make bricks and build a tower to reach the sky (Heaven). But God was displeased and they were punished by being dispersed all over the earth and, in addition, by being given separate languages so they would not be able to understand each other. (See Genesis 11:6 - 8.

2000 B.C.: Abraham – Historical Times
• In about 1900 B.C. Abraham, the family leader of a remote desert tribe, came to Palestine. His descendants became the leaders of the Twelve Tribes of Israel. They moved to Egypt and were enslaved there in about 1700 B.C.

1250 B.C.: The Exodus
• Moses led the Israelites out of Egypt.

Contradictions in the Book of Genesis

Genesis 1 describes human beings as created on the sixth day, as the last act of creation. Land, sky, light, sun, animals, plants, and so on, were all created before human beings. The exact number of humans is not specified.

Genesis 2, the second version of creation, tells of God creating first the universe, then man, followed by the Garden of Eden. After the naming of the animals, woman was created from the rib of the man. Creationists combine these two contradictory versions by different authors into one historical event. The placement of two contradictory stories next to each other by the final editor makes it plain that neither story was intended to be taken literally. There is a similar discrepancy in Genesis 1 where light is created on the *first* day, and the sun, moon and stars on the *fourth* day – although making light first is not inconsistent with the biblical concept of creation from pre-existing material.

These are just two examples. A closer examination of the two versions shows further descrepancies.

Genesis, Chapter 1

In Genesis 1:2 we are told that the earth was formless and desolate; there was a raging ocean in total darkness. This is chaos. The

Semitic mind did not think in abstract terms. There was no concept of "nothing." *Nothing* was a Greek concept taken over by Christian thinkers. For the Semitic mind the end of the world would have been marked by its lapsing back into chaos. The idea of God creating "out of nothing" is not a biblical concept.

Day One. God separated light and darkness; He named the light "Day" and the darkness "Night." This has led some creationists to call for a scientific study of darkness!

Day Two. A dome was made to divide the water. Part of the water was kept above the dome with space underneath the dome. The name given to the dome was "Sky.

Day Three. The land and water were separated, as were the plants and trees. The land was named "Earth" and the water was named "Sea.

Day Four. The sun, moon and stars were made. And here we see a sort of contradiction depending on the translation. If there is no sun or moon till Day Four, how can there have been *day* or *night* for the preceding "days"? At the end of each of the first three days, it is stated: "Evening passed and morning came." Yet there are no "days" until Day Four when the sun is created. In addition, on this day the sun is put there to separate light and darkness, which had already been done on the first day.

Thus it becomes difficult to accept that the intention of the writer was a scientific account, even in terms of the science then current.

Day Five. The waters were filled with many kinds of living beings and the air was filled with birds. And then God commanded that there be animals – domestic and wild, large and small.

Day Six. Human beings were made, both male and female. They were given control over the earth. The writer is definite in saying that Israel's God made all humans. Different gods did not make different nations.

Day Seven. The making of the universe was completed and God declared a day of rest. This is to emphasize the divine sanction accorded to the Law of Moses which decrees the Sabbath as a day of rest.

It is now accepted that the first creation story is derived from the Babylonian myth *Enuma Elish,* named from its opening words, "When on high ..." Seventy years of captivity in Babylon would have given the Jews plenty of time to become more than familiar with this.

A Comparison of *Enuma Elish* and *Genesis*

Enuma Elish	Genesis
An account of the birth of the gods and various conflicts between them	
Divine spirit and cosmic matter are coexistent and coeternal	Divine word creates cosmic matter and exists independently of it
Primeval chaos; Ti'amat, enveloped in darkness	The earth is a desolate waste, with darkness covering the deep (tehom)
Light emanating from the gods	The creation of light and the separation of light and darkness
Marduk's work of creation: a) The creation of the firmament	The creation of the firmament and the dividing of the waters
b) The creation of dry land	The creation of dry land, the sea and plant life
c) The creation of the luminaries	The creation of the luminaries, the creatures of the sea and the birds
d) The creation of man The building and dedication of Esgila, the temple complex	The creation of land animals and human beings; God instructs Adam and Eve and blesses them
The gods rest and celebrate; the hymn to the creator, Marduk	God rests from all his work and sanctifies the seventh day
Epilogue	

Source: Skehan, 1986, p.12.

Genesis, Chapter 2

In Genesis 2:5 we are told that when God made the universe there were no plants or rain on the earth. Then He fashioned a man from the soil and brought him to life by breathing into his nostrils. After making man God made a garden for him to cultivate. And then, because man was lonely, God made animals for him from the earth and woman from the rib of the man.

The second story of creation is most powerful in its use of symbols. Breath is a symbol of life. Earth and breath are elements found in other stories and the tree of life is an archetypal symbol found also in other cultures. The source or sources of the second story are unknown.

Comparing the Two Stories

The two stories are impossible to reconcile logically. They make sense only as different stories used to answer different questions.

The first story answers the question, what is the origin of man? The second story answers the question, why is man the way he is – an erring creature moving to death?

Taken overall, the first eleven chapters describe the origins of the world and the early history of human beings. They underpin the Law of Moses. The literal reading of Genesis used by creationists is a travesty of how to *interpret* the Bible. This is why their understanding of the religious truths inherent in Genesis is as barren as their science.

Modern Biblical Scholarship

Mankind has recorded a common belief, in all ages and in all cultures, that there is some force, power or meaning present in creation. Something transcends, or is beyond, both mankind and creation itself. Common to humanity from before the dawn of history has been a concern with the search for an ultimate meaning to which many names have been given, the most prevalent being that of "god."

The civilizations bordering Israel had their own religions, creation stories and gods. What made the Israelites unique was a firm conviction that their God was superior to other gods and that Israel had been chosen above all other nations to be a light to all nations. The Bible consists of many books by many authors, written at different times and using different genres, some of them entirely unknown in Western literature. In addition, it was written by a Semitic people whose thinking patterns comprised concrete and not abstract ideas.

Genesis, the last book written in what Christians call the Old Testament, expresses an understanding which had emerged at the end of a thousand years: that Yahweh was not merely the covenant God of Israel, but the Creator of the world and of all nations.

While the Bible is a book of faith, it is also a subject of legitimate scientific study with the same legitimate autonomy as any other branch of scientific research. The scientific study of the Bible has proceeded over the past century and yielded abundant knowledge. Biblical research has become an exciting field of study, and we now have insights which were simply not available even a few decades ago.

Archeological and modern biblical scholarship over the past century or more has provided so much information that scholars today know more about the times of the Old Testament than was known in the time of Christ. Between five and six centuries had elapsed between the writing of the last books of the Old Testament and the lifetime of Christ.

Over the past 1800 years or so, and until relatively recent times, Christianity taught not only that God created the world but that the world was created in the way described in Genesis. It was the common belief of Christians that the first five books of the Old Testament were written by Moses. By today's standards, the entire Western world was "fundamentalist."

The choice in recent years has become that of either discarding Genesis as outmoded science or looking more closely to seek the underlying meaning. What follows is an overview of some of the ways in which Genesis is interpreted today. There is ecumenical consensus on the theology of Genesis.

Chapters 1 to 11 recount the story from creation to the birth of Abraham. Most modern Bibles, such as the *Good News Bible* , label these chapters as prehistory. Partly because of this there are some fundamentalists who state that all modern translations of the Bible are the work of Satan. They usually maintain that the Authorized Version, which is the English translation authorized by King James I in 1611, is the only Bible that is truly the Word of God. No one knows what they think of the hundreds of translations into languages other than English.

Some key elements in Genesis 1 to 11 are the first creation story, the second creation story (concerning Adam and Eve), the lists of ancestors, the Flood and the Tower of Babel. These were written in the mindset of the cosmology of the Middle East at that time. This included a flat earth with a dome. The sun, moon and stars moved across the surface of the dome, and floodgates in the dome allowed the rain to come through.

The First Creation Story

The first creation story is based on an older story, written down some time after the return of the Jews from captivity in Babylon in 538 B.C. A possible and not entirely improbable origin for the six-day creation story is as follows:

At some time during the almost three generations of captivity it became apparent to the Jewish elders that the young were being subverted by Babylonian civilization and culture. How could the puny unseen god of an enslaved people compete with the mighty Marduk, present in his temple with the trophies of the God of Israel at his feet? Marduk, after a mighty struggle with evil gods, had created the world and poured his favors on those who acknowledged him.

The older rabbis countered by turning the creation myth of Babylon back upon itself. Several key ideas were added to justify and strengthen the faith of the young people. The unseen god created effortlessly, without any struggle: "Let it be!" The unseen god created everything, both good and evil: "He was pleased with what he had made." He created in six days, thus justifying the Sabbath. He created man and woman equal. This last idea was a revolutionary statement at a time when the market price of a woman was three-quarters that of a camel!

The Second Creation Story

There are currently various interpretations of this complex story. Suffice it for the moment to indicate briefly one interpretation.

We are each of us both Adam and Eve, tempted and tempter. The Garden of Eden is childhood innocence – the innocence we can remember but to which we can never return. The eating of the fruit of the Tree of Knowledge, of good and evil, is the first time we do something we know to be wrong. From this moment, we know good and evil, and must leave the garden of innocence. But also important is that only by leaving the garden of innocence can we know good and develop as human beings.

The Genealogies

The genealogies, or lists of ancestors, occur in Chapters 5, 10 and 11. They give names and life spans which were used by Bishop Ussher to calculate the age of the earth. They are based on the second creation story wherein God created a single pair of humans, and not on the first creation story wherein God created the human race on the sixth day.

For example, Genesis 5, verses 1 through 8:

> This is the list of the descendants of Adam. When God created human beings, he made them like himself. He created them male and female, blessed them, and named them "Mankind." When Adam was 130 years old, he had a son who was like him, and he named him Seth. After that, Adam lived another 800 years. He had other children and died at the age of 930.
> When Seth was 105, he had a son, Enosh, and then lived another 807 years. He had other children and died at the age of 912.

Chapter 5 continues for another eight names and ends with Noah's sons, Shem, Ham and Japheth.

It would seem to be relatively simple to calculate back to the time of creation. However, since the average age of the eleven people in the lists was around 900 years, and because Adam was alive when Noah's father was alive, it must surely indicate that the biblical writer was using a common literary device to show that *the origin of all human beings was from the God of Israel and not from the gods of Israel's neighbors* .

The Flood

The tablets describing the Babylonian myth from which the writers of Genesis derived the story of the Flood were discovered and documented as far back as A.D. 1872. The punishment of human beings by the Flood is followed by the Covenant of the Rainbow, a reassurance from God that never again will such an event happen. Sin - punishment - salvation is a recurrent theme in the Bible.

The Tower of Babel

The biblical author used an ancient story (or stories) explaining the origin of languages, to assert that sin alienates all human society from God and all mankind from one another. Contemporary Greek myths commonly told of gods who saw humans as becoming too powerful and a threat to themselves. There are similar sentiments expressed in the Tower story:

> Then the Lord came down to see the city and the tower which those men had built, and he said, "Now then, these are all one people and they speak one language; this is just the beginning of what they will be able to do. Soon they will be able to do anything they want! Let us go down and mix up their language so that they will not understand each other." So the Lord scattered them all over the earth, and they stopped building the city. The city was called Babylon, because there the Lord mixed up the language of all the people, and from there he scattered them all over the earth. (Genesis 11:5 - 9)

We still use the word "babel" to describe meaningless noise or a noisy group of people. Overall, the myths and stories in the first eleven chapters of Genesis describe the alienation of mankind from God, of man from woman and of men from each other. They set the stage for the entry of historical Abraham and the beginning of the story of salvation in Genesis 12. The creationist reading of Genesis renders it an obsolete science textbook, a curious but irrelevant group of ancient tales of no importance today. This approach ignores the fact that in all religions myth and story are used to convey religious truth.

Chapter 7

How Did We Get Where We Are ?

There are hundreds of publications by creationists which defend the Flood. This number is constantly augmented as anticreationist forces publish new objections to the arguments and evidence brought forth by creationists.

Each new argument put forward by creationists seems to enmesh them deeper and deeper into their own web. In addition, creationists need to call upon divine interventions more and more. For example, when confronted with the difficulty of Noah's collecting all the animals worldwide to place aboard the Ark, creationists reply that two of each species were temporarily endowed by God with a migratory instinct which enabled them to travel to the Ark. When asked how this made it possible for blind moles, earthworms and wombats to swim across the sea or for whales to flop across the land to find the Ark, they again call on the intervention of God. And so it goes on and on.

The creationist defense of the Ark now involves so many interventions by God that it seems likely that if God had known ahead of time of the thousands of miracles demanded by creationists He would never have gone ahead with the project.

They just keep on fabricating the evidence, glossing over the errors, and use the same old publications and films which have proved effective in schools and church audiences.

Yet for creationists the Ark and the Deluge are so important that any number of miracles will be invoked if necessary. Without this appeal to miracles there is no explanation for sedimentary rocks, mountain building, the erosion of the Grand Canyon in America or the Blue Mountains in Australia. Nor is there any explanation for fossils, oil and coal – or even the migration and hibernation of animals. The list goes on forever.

The problems with the Tower of Babel are just as serious. If one reads the Bible the creationist way, God did thousands of miracles from the times of Noah and suddenly stopped when Abraham was born. But the events which creationists proclaim as occurring consequent upon the collapse of the Tower of Babel require as rigorous a defense by creation science as does the Ark.

The Ark and the Tower

It is a key creationist belief that all life on earth, including human, is directly descended from Noah's family and the plants and animals he took aboard the Ark to survive the Great Flood in 2345 B.C.

It is further claimed that human beings who now exist – with their differing civilizations, cultures and languages – date from no earlier than 2195 B.C., approximately 150 years after the Flood, when Noah and his descendants were punished and dispersed for attempting to build a tower to the sky. According to the Bible Noah lived to the ripe old age of 950. He died 350 years after the Flood.

Near Neighbor Problems – Egypt

From Roman times the pyramids of Giza, including the great pyramid with some 2.5 million massive stone blocks, have inspired a sense of awe and wonder. The great temples along the Nile are also monuments to a civilization which had developed quite extraordinary skills.

Stone age cultures existed along the Nile valley at least 20,000 years ago. The great pyramid tombs at Giza were built between 2686 B.C. and 2496 B.C. The Egyptians kept detailed records and the dates of the various dynasties are known with great accuracy (*Lexicon*, 1987, vol. 7, p. 81).

The great pyramids can be dated at least three centuries before

2195 B.C., the year in which creationists claim The Tower of Babel was built. While this contradicts creationist chronology, there are further implications which are even more unlikely.

Even if Noah's progeny had increased prolifically to 100,000 in the 150 years or so after the Flood, it is beyond comprehension to claim that a portion of these people migrated to Egypt and immediately decided for no apparent reason to spend all their energies on building pyramids in which they buried their dead pharoahs together with their treasures.

There are people who believe the pyramids were built by aliens from space. This is a much more likely explanation than that of creation science.

None of the above beliefs seems to deter creationist biblical scholar A. J. M. Osgood (1986b, pp. 77 - 87). This is the same Osgood who found a fossil orange. In his article "The Times of Abraham," he proceeds to slice a thousand years out of Egyptian history, as well as numerous pharoahs, because they don't fit biblical chronologies.

Other Neighbors

Other well-studied cultures such as India and China require much more than a thousand years sliced out. This is because they would have had to migrate from the Middle East in 2200 B.C., develop cultures based on a different language which they had suddenly received, as well as set up the appearance of being in their new lands for some thousands of years. This does not take into consideration the changes in stature, skin coloration and facial features which, according to artifacts, have been present since the beginning of recorded history.

In another article, "A Better Model for the Stone Age," Osgood (1986a, p. 90) dates the stone age, by careful study of the Bible, as 2300 to 1870 B.C., which corresponds to the time of Abraham. This is to fit in with his own Flood chronology. In doing so, he seems to forget that the biblical Abraham was described as a nomad and not a stone age man.

Osgood's view appears to contradict archaeological evidence that the stone age in Europe finished around 10,000 B.C., when farming and towns came into being. It is easy to figure out that this is about 8000 years before Osgood's date! Carved stone figurines of the Earth Mother have been found in northern Europe. These have been dated as 30,000 B.C. and earlier.

Continental Drift

The first modern atlas of the world was published toward the end of the sixteenth century. Since that time many people have wondered why the continents seem to fit like a jigsaw puzzle. It was proposed in the early twentieth century that the continents had indeed once been one large landmass. However, the theory that the continents had drifted apart from one another was not fully accepted until some fifty years later. Evidence since, from a variety of sources, supports continental drift by the mechanism of moving crustal plates (Ritchie, 1987, p. 63).

The drifting apart of the continents explains why plant and animal life on the different continents is similar but different. Satellite measurements show that the continents are still moving apart at a rate of several centimeters a year. The "ring of fire" around the Pacific is caused by the grinding together of the separate "plates," which support the continents and ocean floors. Along the lines where the plates meet are zones of earthquakes and volcanic activity. Most recent encyclopedias show the details.

When the continents were joined, there was no barrier to plant and animal life spreading across the single large landmass which then existed. However, the separation which has been in existence for the past 100 million or 200 million years accounts for the enormous varieties of plants, mammals, birds and insects which are unique to each continent. While their common ancestry dates back to one large continent, their different paths of evolution date back to the time of separation of the present continents.

Quite apart from the thousands of volumes of scientific evidence in support of continental drift, no one has yet thought of an alternative theory which makes such good sense in explaining the life-forms that are similar but different in the various now-separated parts of the world.

Having said the above, I find it near painful to write the explanation that creationists offer:

The unique flora and fauna of each continent occurred after

all the plants and animals were released from Noah's Ark about 4300 years ago. Kangaroos, koalas and wombats migrated through southeastern Asia and crossed the seas to Australia, not going anywhere else. Eucalyptuses floated across to Australia and nowhere else. Emus swam across the Torres Strait to Australia and nowhere else. On arrival they joined up with their fossil ancestors which are found in Australia and nowhere else. (Price, 1987, p. 21)

The list goes on and on.

One could, of course, apply the same reasoning to the llamas and armadillos of South America, but the point has been made.

Remember also that only a year or two before, according to creationists, Noah's family had visited all the different continents in order to collect living samples of every species to take aboard the Ark.

Common ancestry can be readily shown by analysis of DNA sequences. For example, geologists date the separation of Africa from South America as 80 million years ago. Charles Sibley compared the DNA sequences of African birds with their similar counterparts in South America. The percentage differences in the DNA sequences confirm the geological time estimate (Jukes, 1988).

Creation Science Answers Back

As always, creation scientists have come up with a most ingenious way of countering the above arguments. They claim that one of their scientists, Barry Setterfield, has proved that continental drift occurred *after* the Flood. Moreover this amazing discovery is the work of an Australian creationist. For this he has received accolades and praise from American creationists.

The Australian Export Developments Board has shown the appreciation of the Australian taxpayer by giving the Creation Science Foundation a grant of $24,693 to promote Setterfield's book overseas. Australia's greatest writers have seldom, if ever, received a government grant of such magnitude to help them along their way. But none of them has been able to slow down the speed of light. The legality of the grant itself has been questioned:

The Export Development Grants Board did not ask the

Attorney-General's Department for an opinion as to the constitutionality of the grants. The Board refused to make public details of the CSF's submissions on the grounds that it would not reveal information about an applicant's affairs. However Dr C. Wieland, President of the Creation Science Association Inc (sic), said "two of their people went to the United States to promote the sales of that (Setterfield's) book. That was regarded by the government, and quite correctly so, as a straight commercial undertaking, which benefits Australia as far as export revenue was concerned." (Bridgstock, 1987, p. 83)

Barry Setterfield claims to have proved that the speed of light has slowed down so dramatically over the past 6000 years that events thought by evolutionists to have taken place millions of years ago really took place a few thousand years ago. Hence the animals did not have to swim across oceans on their journey to or from Noah's Ark because at that time all the continents were joined.

Setterfield's claim that the continents rapidly separated shortly after the Flood solves one problem but presents a multitude of others. The drifting apart of the continents at the present rate of 2 or 3 centimeters per year is the cause of the earthquakes and the 500 volcanoes which are active today.

If the "drift" from one giant southern continent to the present positions of the continents had occurred in the space of a few years, the surface of the planet would have reverted to molten lava. If, on the other hand, the drift had occurred over the past 4500 years (Setterfield doesn't specify an exact date), then the average rate of drift would have been about 4 kilometers per year. Forgetting for the moment that this speed would be observable, the consequences are almost as drastic as if it happened in a few months.

Plimer (1988a) calculates that if Setterfield is correct then over the past 4500 years there would have been a catastrophic earthquake every 6 minutes, a tidal wave every 18 minutes, a major volcanic eruption every 12 minutes. In addition, the sun would be blocked out completely by the dust, and air temperature would fall to -38°C (-36°F). As if that were not enough, the oceans would be boiling. There is no record in Greek or Chinese history of such a catastrophe which would have lasted thousands of years. This is not surprising since under those conditions there would have been no Greek or Chinese history. Nor any other sort of history.

But one thing is certain. Creation science always finds a way out of it, even if they have to contact their god of the gaps and cajole him into becoming a god of the chasms (Chapter 9 discusses in detail the nonsensical consequences of Setterfield's theory). As well, the way in which the creation scientists at the ICR admit Setterfield's error, while at the same time keeping his book on sale, illustrates a devious ingenuity far beyond the ken of the normal mode of scientific thinking.

Australia

Archeological finds indicate that the Aborigines have been in Australia at least 40,000 years. It may be that Australia is the site where modern man first evolved. Eugene Stockton *et al.* (1987, p. 78) report that a dig on the Nepean River near Cranebrook, New South Wales, shows continuous occupation by a single community for 45,000 years. This, they claim, is the longest occupation of a single living-place in the history of mankind. They further claim that the emergence of modern man in Australia may predate modern man in Europe by as much as 12,000 years.

Whatever the final consensus of science may be concerning the presence of Aborigines in Australia, it is certain that they have been here 40,000 years at the very least. This is tens of thousands of years longer than the 4000 years claimed by creation science.

It is an assault on the mind to claim that the Aboriginal peoples wandered back to Australia and quickly dispersed to occupy the whole continent so that by the time of the first contact with Europeans in the 1500s there were hundreds of tribes, each with its own dialect and own territorial boundaries. Yet this *is* what creationists claim. Systematically and continually they provide "scientific proof" in their journals and films.

The following statement from *Ex Nihilo Technical Journal* speaks for itself:

For ten years now our researchers have collected, collated and compiled the fast-disappearing evidence from the

Aboriginal cultures of Australia. Now we have one of the most extensive collections of Aboriginal myths and legends in the world. The result? The research shows conclusively that the Aborigines are not and never have been evolutionist. It also demonstrates they have not been in this land for 40,000 or more years, that they came here after the time of Noah's Flood and brought with them their memories of Creation and the Tower of Babel. These memories are vividly preserved in their Dreamtime stories. This research is proving invaluable in repulsing the lie of evolution which denigrates the Aboriginals (sic) to being people who still have yet to catch up in the evolutionary race. With your financial support, these results can be fully published, not only for European races to read, but perhaps more needfully for the Aborigines themselves, before it is too late. (1986, p. 155)

One can only wonder how Aborigines feel about this. There are many who see Aboriginal spirituality, which places man and the land in intimate harmony, as being of great relevance for the white man in understanding and caring for this timeless land and ensuring a long-delayed justice for its original owners.

Surely it must be close to blasphemy to dismiss the aspirations, hopes and religious history of a proud people as racial "memories of Creation and the Tower of Babel."

Chapter 8

Creation Science and Dinosaurs

It is somewhat humiliating for scientists, and others who spend a great deal of time exposing the errors and illogicalities of creation science, to find that creationists see dinosaurs as a much greater threat than the ranks of scientists arrayed against them.

The love affair between creationists and dinosaurs began in the years of the Depression. The inhabitants of Glen Rose, Texas, found dinosaur tracks in the beds of limestone excavated in their own backyards. These were sold to tourists in curio shops. When the supply dwindled the locals carved more, including some human tracks.

Paleontologist Roland T. Bird from the American Museum of Natural History saw the tracks for sale in the curio shops and traced their source to the Paluxy River in Texas. His account of the dinosaur tracks was published in 1939 in *Natural History* .

These were the days before the foundation of the Institute for Creation Research. A group of "special creationists," as they were called then, including Clifford L. Burdick, searched out the fakes that Bird had rejected, and declared them genuine (Godfrey, 1981, p. 27). The "footprints" discarded by Bird included tracks of single dinosaur toes which resembled human footprints when looked at back to front, partial prints made as dinosaurs touched the mud while swimming, natural erosion of the rock as well as other tracks carved by the locals.

Burdick attacked Bird for concealing evidence and the next forty years saw Paluxy become a war zone between evolutionists and creationists. For creationists Paluxy has now achieved the status of a religious shrine.

Survival at Paluxy

An intriguing aspect is not so much that the alleged footprints are in Cretaceous limestone 200 million years old, since creationists automatically deny the existence of rocks older than 6000 years, but that the depth of the limestone sediments beneath the footprints is more than 2 kilometers (around a mile and a quarter). It is basic to creationist geology that all deep sediments were laid down during the year of the Great Flood. This includes the 50-kilometer-thick (c. 30 miles), fossil-bearing sediments which comprise Mount Everest and the Himalayas. Otherwise they would be forced to admit a very great age for the earth.

The Himalayas are tilted in some places 30 degrees or more to the horizontal. Since sedimentary formations by definition are laid down horizontally they must, according to creationist geology, have tilted over the past 4000 years. One would expect, because of this, to find leaning monasteries in Tibet. There are none. The scientific explanation is that the Himalayas are at the junction of two crustal plates. But creationists don't believe in continental drift.

The mind boggles at the thought of an isolated group of humans and dinosaurs at Paluxy treading water for a year in conditions which laid down 2 kilometers (over a mile) of sediment. Somehow during the deluge they managed to keep together while the water rose more than 8 kilometers (more than 4 miles) above sea level, covering even Mount Everest, if it was there! As the water subsided the herd of dinosaurs with their human companions touched down at Paluxy after starving and treading water for a year. They immediately started walking around in the mud, presumably looking for food. Overlooked in this scenario is that the biblical account is quite clear that only Noah, his family and the plants and animals on the Ark survived the deluge, and according to Henry Morris the Bible tells no lies.

When queried on this, John D. Morris, son of Henry, indicated a high plateau 40 kilometers (25 miles) distant as a place which the only known survivors outside the Ark could have used as a refuge. Nothing daunts a born creationist.

The same John Morris has argued that the winds during the Flood made a tornado seem like a gentle zephyr. The wind-blown waves must have reached a height of a hundred meters during the Flood, he says. These waves and winds would certainly have washed away our little contingent of survivors. In addition, the height of the plateau on which the group ·sheltered is several kilometers (a few miles) lower than Mount Everest. There is no record of Morris's reply to these objections but the answer is obvious. The creationist god of the gaps equipped both man and dinosaur with anchors and snorkels.

In January 1986 John Morris told a conference at the University of New South Wales that he had "encouraged the creationists to refrain from using the Paluxy as an anti-evolutionary argument" (Morris, 1986, p. 722). Thomas Jukes reports that

> the fake exhibit at the Institute for Creation Research, showing an 80-million-year-old human being beside a dinosaur, has been "closed for revision", and that the "casts" have been removed. (Jukes, 1986a, p. 722)

What Morris said was that only *those* footprints found so far were in doubt.

Staff at the Creation Evidences Museum at Paluxy have now found a tooth among the dinosaur footprints. Ten creationist dentists assert that the tooth belonged to a child. Paleontologists disagree. The tooth finders claim this is out of spite. Now the search is on again with renewed enthusiasm and the backing of the dentists. Sooner or later a tooth will be found with a filling in it.

Tiptoe Through the Trilobites

William J. Meister and the same Clifford L. Burdick of Paluxy fame have since found prints made by boots, sandals and a barefoot child while hunting for fossils near Antelope Springs, Utah. These footprints were found among trilobite fossils in Cambrian rocks about 600 million years old. Trilobites are small shellfish-like animals which existed in the seas 400 million years before the dinosaurs – long before life on land. This was disclosed on 1 March 1973 when Duane Gish and Reverend Boswell comprised the creationist team in a debate at California State University in Sacramento (Conrad, 1981, p. 30).

The New-Look Dinosaurs

Creationists have recently stepped up their campaign of films, texts, children's books, toys, exhibits and posters which show humans cohabiting with dinosaurs. This is in response to the startling new finds on dinosaurs which have occurred in the past two decades. The image of dinosaurs as massive, dimwitted, slow-moving reptilian creatures with miniature brains is now gone. This traditional view was turned around in the early 1970s.

Newly discovered breeding grounds in Montana show that dinosaurs had communal nests and that after hatching, the babies were cared for in the nest for up to a year. Herbivorous dinosaurs traveled in packs of at least twenty-five and the young were protected by being placed in the center of the herd while traveling or grazing.

There were hundreds of different types of dinosaurs. They ranged from small ones the size of a chicken to those which reached a length of 30 meters (about 98 feet). Their weight ranged from a few kilograms (or several pounds) to over 70 tons. Most were herbivores but some of them were among the most ferocious meat-eating predators ever to have walked the planet.

Dinosaurs were reptiles but were warm-blooded, as are both mammals and birds. Their activity level was comparable to that of birds and their nesting habits have been likened to those of large birds such as emus. They traveled on their hind legs and analyses of their tracks indicate in some cases speeds up to 80 kilometers (50 miles) per hour.

New analytic methods show that dinosaurs did not have

exceedingly small brains. The ratio of brain size to body weight for most dinosaurs is the same as that for living crododilians, and some dinosaurs were at least as complex behaviorally as crocodilians and birds. Modern crocodilians have a wide behavioral repertoire, including advanced maternal and paternal care. (Chure, 1984, p. 47)

The new-look dinosaurs became popular in 1976 with the best-seller by A. J. Desmond, *The Hot Blooded Dinosaurs: A Revolution in Paleontology*. Since then they have become a fad to which creationists have strongly reacted.

The Creationist Response

In response to the dinosaur craze sweeping the United States creationist Paul Taylor expresses his deep concern about dinosaurs. Taylor is production director of Films for Christ and through his films has a vested interest in creationist claims that man and dinosaurs cohabited in recent times. He urges fellow creationists to rally to the cause:

They're everywhere! Dinosaurs of every size and color. In America you can see them menacing shoppers from store windows, leering from greeting cards, dangling from key chains and splashed across T shirts. Dinosaurs! And kids love them. *But that's the danger*! (Taylor, 1987, p. 23)

Creationists have for some time been unduly obsessed with dinosaurs and Taylor's article gives some of the reasons:

Teachers are generally encouraged to use dinosaurs to promote learning skills. Dinosaurs have become almost indispensable as teaching aids. Educational supply stores are usually well stocked with such supplies used in teaching subjects as varied as English and reading, arithmetic and history not to mention science. All along the way, dinosaurs are related to evolutionary concepts that lead children away from the truth revealed by God in His Word.
This evolutionary indoctrination is thorough and reinforced.

After years of exposure, beliefs become firmly established in people's minds. Eventually, for most, even to look at a dinosaur automatically brings forth mental images of evolution – ideas effectively re-emphasized in the home by hours of television documentaries, beautiful *National Geographic* specials and evolutionary articles. (p. 25)

He sees dinosaurs as a major temptation for Christians:

Often, even Christians cannot withstand the temptation to believe. Christians by the millions have been deluded, and their faith in the Bible damaged, owing to the misleading information they have been taught about dinosaurs. (p. 25)

Dinosaurs are probably the most believable of fossils. There is nothing alive on our planet today that remotely resembles them in size and variety. Children readily accept that they are millions of years old. Dinosaurs, for creationists, are a subtle attempt by evolutionists using unlimited funds to indoctrinate the young in favor of evolution.

Taylor objects to dinosaur parks because public money is being used to fund evolution:

Even tax money goes to support dinosaur exhibits in museums and parks and publications. And children willingly and eagerly listen to the indoctrination because they are fascinated by dinosaurs.

As Dr. Henry Morris explains, "It is primarily the dinosaurs which have caused many people to think that the fossil world is so different from the modern world that it must have existed millions of years ago, long before modern plants and animals evolved."

The results can be tragic. Students frequently cannot accept the Bible because of their beliefs about evolution (and dinosaurs). And if they can't accept the Bible they cannot accept Jesus Christ. This fact is reported daily by ministries working on college and university campuses. In addition, these same deluded young people will grow up to be the leaders, educators, and politicians of the next generation. (p. 26)

Taylor continues in the article to give a comprehensive list of ways in which to prevent the young being seduced by dinosaurs. He says children should be provided with creationist books, movies, videos, audio tapes and activity sheets so they will learn at a young age that dinosaurs and humans were contemporary with each other. He says children's natural passion for dinosaurs can be used to turn children to their Creator by learning the *true* history of the planet. Parents should ensure that libraries are stocked with dinosaur books (presumably creationist in theme) and that schools are encouraged to add dinosaurs to the curriculum as well as to the school bulletin boards. Kids should be told about dinosaurs in church. Families should attend "Dinosaur Nights." Such actions he sees as critical "especially during the current dinosaur fad."

Taylor is blissfully unaware of the implications for the young of making the presence of dinosaurs on the Ark 4000 years ago a precondition for the acceptance of Jesus Christ. No wonder creation science leads to atheism.

Australian Dinosaurs

Australian creationists have shown great interest in the recent discoveries of dinosaur fossils in Queensland. Their visits to these places are no doubt motivated by the thrill of being the first to discover human footprints among the Australian dinosaur tracks. With any luck the finds could be sufficient to totally demolish evolutionary theory. Another of Osgood's oranges would do the trick. Even one of Snelling's gold chains or a fossil hat would be welcome.

Any finds of this nature would take pride of place at the proposed creationist display center in Brisbane.

The long-running affair between creationism and dinosaurs illustrates the long-term cycle of action and reaction that is part of the creationist psyche. First there is the stage of exhilaration, a new find which demolishes evolution. Second is the publication and propagation linked to renewed evangelism. Then comes the time of doubt. Like a rusty bucket full of water, more and more leaks appear with use. The leaks are plugged not by purchasing a new bucket but by closing the mind tighter. Now it's time to invoke Satan of the gaps. All contrary evidence is labeled the work of the devil because scientific evolution is a tool of Satan. Thus the mind is finally proofed against the danger of any new idea. The tiniest cracks are sealed. Action is now taken to ensure that the minds of their children are never affected by their own doubts about dinosaurs and evolution. More and more materials are produced and "lying in the name of God" is justified.

Dinosaurs Aboard the Ark

Creationists claim that dinosaurs, like all other animals, were instantly created at the beginning of the world. However, these same creationists don't explain what the meat-eating dinosaurs lived on in the early years of creation before there was pain or death. Cows were contented with their munching teeth and four-chambered stomachs. But what about lions and tigers? They had a problem. They would have had to wait anxiously for the first sin so they could get a decent meal.

Some creationists explain how these huge animals all fitted into the Ark by saying that Noah took them aboard as eggs. In Genesis there is no mention of taking aboard anything but animals, although it is generally supposed that plants were included. In any event freshly hatched dinosaurs are very much in need of parental care. Nor does this explain how Noah was able to identify male and female eggs.

The point is that creationism is tangled up in its own web. The more perceptive creationists recognize dinosaurs as an Achilles heel. The original Achilles was safe so long as it was secret that he could be killed only through his heel.

In summary, to have dinosaurs (and other animals) on the Ark, there is at the very least a fivefold problem for creationists:

1. Traveling to all parts of the world to capture all species

which fossil evidence shows were present in that part of the world before the Flood. This includes Antarctica.

2. Getting all these safely to the Ark.

3. Fitting all known species of animal and plant life that exist or have ever existed into the Ark.

4. Feeding, cleaning and watering them during the journey, as well as providing provisions for the twelve months' stay on the Ark. This includes keeping the meat eaters fed.

5. Staying afloat in unimaginable seas and winds.

It is perhaps interesting to examine this last point briefly. The longest wooden ship ever built was the six-masted USS *Wyoming,* with a length of 83 meters (329 feet). In heavy weather the hull twisted like a snake and required diagonal steel strapping to stabilize it. The ship was unsafe in deep water and could be used only along the coast. The Ark was half again as long, at 137 meters (450 feet, 300 cubits). After nine six-masted vessels were launched between 1900 and 1909, all having similar problems to the *Wyoming,* it was concluded that 90 meters (295 feet) was the absolute maximum length for a wooden ship (Moore, 1983, p. 4).

This is *some* problem, and the fact that dinosaurs became extinct about 65 million years ago seems a minor point in comparison.

Chapter 9

Slow Down the Universe – I Want to Get Off

> *The speed of light* is the most accurately known of the universal *constants*. These do not vary in space, place or time. They are the "fingerprints" of our universe. They are simply there. No one knows why they have the values they do. Thus light has a speed of 300,000 kilometers per second (186,000 miles per second). Other such fundamental constants are those which occur in electricity, magnetism, the mass of protons and the nuclear force. They are interconnected and the smallest changes in any one of these fundamental constants would drastically change the structure of the physical world (Davies, 1983, p. 188).
> Creationist research has changed the speed of light in order to show that the universe is no older than 6000 years. The effects of changing the value of a universal constant such as light are shown to be an impossible fantasy which bears no resemblance to our universe.

Viewers of the NASA moon landings that were shown live on television noticed a delay of several seconds in conversations between mission control and the astronauts. The delay lengthened as the probe ventured farther from the earth toward the moon. Thus the watching world received a dramatic illustration of the speed of electromagnetic radiation in the form of radio waves.

Scientists know that electromagnetic radiation in the form of light takes about eight minutes to reach the earth from the sun. That is, if the sun were "switched off," we would not know until eight minutes later.

Astronomers with very sophisticated instruments have shown that Andromeda, our nearest neighboring galaxy, is 2.2 million light-years away. This means that light from Andromeda has taken 2.2 million years to reach us. A light-second is 300,000 kilometers (186,000 miles), while a light-year is about 10 billion billion kilometers (6 billion billion miles).

This, of course, causes problems for creationists, who *must* hold to the view that no star can be more than 6000 light-years away. According to creation science, no stars existed before the creation of the universe 6000 years ago. Hence light cannot have been traveling through space for longer than 6000 years.

The Old Solution

Henry Morris, the father of creation science, states that there is really no great problem because the universe was created in 4000 B.C. with an "appearance of age."

> Whatever is truly created – that is, called instantly into existence out of nothing – must certainly look as though it had been right there prior to its creation. Thus it has an appearance of age. (Morris, 1972, p. 62)

Other methods used to date the earth and the universe have been under constant attack in creationist literature. These include carbon and other radiometric dating, the fossil record and nuclear chronometers, which depend on the observed amounts of various isotopes in stars. Nuclear chronometers give an age of the universe in excess of 10 billion years. The results are independent of the speed of light (Selkirk and Burrows, 1987, p. 45).

However, since 1983, creationists no longer find it necessary to expend time and energy attacking the validity of the various methods used to date the universe. Creationist researchers claim a major scientific breakthrough which allows them to accept the evidence from astronomy, radiometric dating, carbon dating, and so on. In so doing they can still prove the universe began 6000 years ago and not 10 billion years ago. This solution represents a major change in the direction of creation science, and is remarkably simple:

Slow down the speed of light !

The New Solution

Any competent mathematician can construct an equation to describe a line joining several points. It does not require a Nobel Laureate to make an equation to fit the tracks of an alcoholic lizard wandering erratically over sand dunes.

Barry Setterfield, Australian creationist and competent mathematician, has analyzed measurements of the speed of light from experiments done between 1675 and 1928. He has derived a theory which gives the speed of light at the time of creation as 200 billion times faster than it is at present (1984a, p. 82). He has also found that, with this new value, all current measurements of the time light takes to reach the earth from distant stars and galaxies will fit comfortably into a time frame of 6000 years, not billions of years as astronomers claim. He adds that the speed of light stopped slowing down in 1960.

Carbon Dating

It is of interest that for years, creationists refused to accept carbon or other radiometric dating as accurate because such methods give results indicating an age greater than 6000 years, and *nothing* can be older than that.

However, things are different now. Using his new theory Setterfield (1986, pp. 171 - 4) now accepts the results of carbon dating, which he has "discovered" by analyzing measurements taken over the past 400 years. Setterfield claims that carbon dating is valid but that the answers it gives are simply far too large because scientists have not allowed for the changing speed of light. After he has corrected this simple error made by not-too-bright scientists, he finds that carbon dating agrees with the Bible that the earth is only 6000 years old. This may come as no surprise to readers.

For example, Setterfield (1986, p. 174) finds that bones and tusks of mammoths from Colorado and Wyoming, carbon dated at 11,200 years old, have a "real" date of 2810 B.C. according to his special theory. This date is still prior to the Flood, but he proposes to refine the model. He has only to lop off about another 400 years to fit in with the creationist date of the Flood (2345 B.C.).

Rejected by Science?

Setterfield's theory on the decay or slowing down of light, if true, would certainly be worthy of a whole box of Nobel medals. But his theory has never been accepted for publication by any reputable scientific journal, which means that his work has not stood up to the scrutiny of scientists. Although Einstein's theories were even more startling when first proposed, scrutiny by his peers allowed publication in reputable scientific journals.

Setterfield's theory languishes between the pages of creationist literature. Creation science accepts it as a breakthrough. While Setterfield is remarkably clever in fending off criticisms, being clever is not necessarily a sign of truth. Even if the mathematics are faultless, the conclusion must be wrong if the premises are wrong and Setterfield's mathematics are certainly not faultless.

A Basic Error

Einstein's mass-energy equation, $E = mc^2$, means that because the speed of light, c, is a very large number, and even larger when squared, a very small amount of matter (mass m) gives an extraordinarily high amount of energy, E, under the right conditions. Conversion of mass into energy is the source of the energy in an atomic bomb, and also of the energy from the sun. Matter is jellied energy and even when we obtain energy by non-nuclear means, such as rubbing the hands together or lighting a match, there is an infinitesimal amount of matter converted to energy.

However, if the speed of light was once 200 billion times faster than it is now, the energy from nuclear reactions would be unimaginably greater than it is now, since the energy depends on the speed of light squared. In numerical notation:

$$200,000,000,000^2 = 40,000,000,000,000,000,000,000$$

Some of the world's richest and most extensive uranium deposits are found in northern Australia. A Geiger counter registers the presence of uranium because the burst of radiation from the fission of a single atom of uranium causes a single click on the counter. Close by a deposit of uranium the clicks can become a nearly continuous note. In the ground the uranium atoms are diluted by rocks and the radiation is harmless. When the ore is treated to obtain a concentrate this is not the case.

However, Setterfield atoms would do more than register a single click on a Geiger counter. Each atom would be equivalent to a miniature atomic bomb weighing 150 grams (5.3 ounces), which is equivalent to the destructive "power" contained in the tactical nuclear weapons deployed around Europe.

In the Setterfield theory, the northern part of Australia and a large part of the rest of the world would have seen millions upon millions of small atomic bombs bursting every second. The dust would have obscured the sun, and creation 6000 years ago would have begun with a nuclear winter.

However, the nuclear winter would have been of academic interest only. The fusion reaction by which the sun produces light and heat would have been enough to incinerate the earth or even to melt it. If this didn't do the trick, the radioactivity present at the center of the earth surely would. The earth itself would become a tiny sun. Even if the Setterfield speed of light persisted for only a short time, the effects would be noticeable now. The planet would not have even begun to cool down.

Apart from all this and presuming the unlikely event that our sturdy ancestors could have survived continuous atomic bomb blasts, a nuclear winter and a molten planet, lighting a match would be equivalent to a small nuclear detonation.

It would seem clear that the first human beings, 6000 years ago, must have had considerable problems. Even touching each other to show affection would be extremely hazardous!

Another Error

An increase in the speed of light would have effects on the nature of light itself. A violinist obtains higher pitched notes by making the string vibrate faster or by increasing the frequency of the vibration. This can be done either by using a thinner string or by using a shorter string, or both. The same technique applies for all stringed instruments.

With light, the situation is somewhat similar, except that it is the frequency alone which determines the energy of the light particles. Increasing the *frequency* of light increases the *energy* of light. X-rays, microwaves, infra-red, AM and FM radio waves, as well as all visible light, are all forms of electromagnetic radiation. They all travel at the same speed. X-rays, for example, have more energy than red, blue or green light because they are at a higher frequency. X-rays can also penetrate the body.

The energy of light is directly proportional to the frequency. If you double the frequency, you double the energy. Setterfield (1984a, p. 71) states correctly in the introduction to his paper that "Light is emitted at a constant wavelength with the frequency proportional to C." (C is the speed of light.)

If we use Setterfield's own data, this means that 6000 years ago, since he states that light traveled 200 billion times faster than the present speed, human beings would have been able to see quite satisfactorily because the wavelength was unchanged. But the light itself, because of its higher frequency, would have an energy comparable to that of very energetic gamma rays. Radiation sickness would occur within minutes and death within hours. No form of life would be possible.

And Yet Another Error

There are certain basic values, or factors, in our cosmos which do not change. So they are called *constants*; they are simply there and must be taken *as they are*. These cosmological constants are embedded deep within the fabric of the universe. It is as though they are the fingerprints which define our universe as unique. They are also interconnected.

As previously stated, the most accurately measured of these constants is the speed of electromagnetic radiation, of which light is an example and for which the symbol always used is C . Other such constants occur, for example, in magnetism, gravity and electricity. These fundamental constants have the "same numerical value everywhere in the universe and at all moments in time" (Davies, 1983, p. 187). Earlier in this chapter we described the sorts of problems which occur when one of these constants, namely the speed of light, is changed. A minute change in the values of these fundamental constants will change the structure of the physical world.

While the mind can construct an infinite number of hypothetical universes with varying fundamental constants, to the best of our knowledge the universe in which we live is the only *possible* universe. Another point of view, however, is to postulate a near infinite number of attempts at creating universes. But certitude on such a question is not possible at the present stage of scientific knowledge and perhaps this state of uncertainty may always remain with us in the future. As Davies says, "the possible existence of other universes remains just as much a matter of faith as belief in God" (1983, p. 189).

A Way Out?

All this brings us back to the speed of light. There is a mathematical escape path for avoiding the gamma radiation which follows when creationists theorize that light has slowed down. We could alter Planck's constant, another fundamental constant which defines, among other things, the smallest limit of measurement.

The energy of light depends not only upon its frequency, which Setterfield claims was 200 billion times faster at the time of creation than now, but also upon Planck's constant. Hence a way to avoid destructive gamma radiation would be to reduce the value of Planck's constant by the Setterfield factor of 200 billion. This would enable the first human beings to have daylight as we know it.

Unfortunately, the reduction of Planck's constant by such a massive amount would mean that Newtonian physics would apply at the atomic level, that atomic and quantum physics would be obsolete and that we would still be left with our nonexistent fantasy-universe.

To make it even worse, if such a universe was possible the mass-energy relationship discussed earlier would still apply. Lighting a fire would still result in a detonation equivalent to a small-scale nuclear war.

The moral of the story is, you can't fiddle with the fundamental constants and still expect to remain in our universe. But someone forgot to tell that to creation scientists.

Deceit With Honesty

When creationists are arguing in front of an audience and they are shown to be palpably wrong, a favorite technique is to admit they are wrong and then offer to retract the statement. Then they leave it in print or else repeat it at the next debate in front of a different audience, knowing they will probably get away with it. If they don't, it's easy to retract it again.

Gish has done this with *Brainwashed* several times. In 1982 he even denied that he wrote it. But sales still continue under his authorship.

When Setterfield published his original papers on the slowing down of light, his mathematics were discredited by a number of scientists both in Australia and in America. Because he always managed to confuse the issue, creationists felt secure that the final evidence against evolution was now in their grasp. Those among creationists who had doubts were conspicuous by their silence. While the issue stayed in the realm of mathematics and was confined to writings in a small number of journals, there was little chance of rebuttal in a way the general public could understand. Setterfield was proclaimed a scientific genius in creation science publications. A grant of $24,693 from the Australian government was an added bonus (Bridgstock, 1987, p. 83).

The status quo changed on the evening of 18 March 1988, when Plimer used some of the arguments detailed previously in devastating fashion against Gish. It became painfully obvious even to the committed creationist audience that Setterfield's theory was nonsense. Gish, in reply, did not attempt to defend Setterfield in the slightest but said that he had always had some doubts about it himself.

Because the debate was filmed by national television and likely to be viewed by an international audience, a new technique was required apart from a one-off retraction/nonretraction statement. The arguments used by Plimer were first published in December 1987, four months prior to the debate, in the forerunner to this book, *The Bumbling, Stumbling, Crumbling Theory of Creation Science*.

Two months after the Gish/Plimer debate the Institute for Creation Research had fabricated a plug for their now leaking theory that light had slowed down. It is an unusual plug to be used when someone is watching but pulled out when nobody is around. It is a plug which avoids public retraction, or even retraction in their own journals. It still allows the Setterfield books to remain on sale and thus continue to provide powerful evidence to the faithful that the universe is only 6000 years old. When found out, they feel like real scientists because they too know how to retract.

In May 1988 the ICR produced a article by Gerald E. Aardsma, "Has the Speed of Light Slowed Down?" Aardsma is head of the Astro/Geophysics Department in the ICR Graduate School – an imposing title. This article rehashes a portion of the published evidence against Setterfield's mathematics. Nowhere in the text or references is there any mention of other published evidence; nor is there mention of anyone except Setterfield. Such an omission is at best extraordinary negligence, and at worst ethically unacceptable. But there is more.

Aardsma (1988, p. iv) states in the conclusion that creation scientists have been subjecting Setterfield's conclusions to "careful scrutiny" since publication in August 1987 and that the results will "soon be available to the creationist community." He urges caution in the interim.

The last paragraph of the article boggles the mind. Aardsma concludes that the entire matter of light from the stars will never be adequately understood because "God has never commanded us to understand all of his infinite works ... but simply to trust Him." In one small leaflet a significant person in the ICR Graduate School has demonstrated pseudo-science, pseudo-religion, pseudo-retraction and deceit by omission.

Now, we wonder, what do you call a technique such as that?

Chapter 10

Some Laws Are Made Not To Be Broken

enry Morris states that the second creation story in Genesis was written down by Adam, who was an eyewitness, and edited by Moses (1974, pp. 205 - 15). He argues, quite logically, that Adam was not alive to witness the details of the creation of the universe outlined in Genesis 1. Morris states that Genesis 1 was probably written down by God in the same way He wrote ten commandments for Moses – etching them on slabs of stone with His finger. Morris sees no problem in Adam stumbling out of the Garden with great slabs of stone on his shoulder. Perhaps Eve helped him perform this task which is neither recorded nor hinted at in the Bible.

The above is but a byplay for Morris. He is more fascinated with the scientific details in God's own words written on the tablets presumably lost by Moses while he was editing. What an archeological find it would be to dig up stone tablets written by the finger of God. Since most creationists assert that the King James Authorized Version is the only accurate text, then presumably Moses may have had to send them away to experts for translation from English.

All this is easy to accept compared to Morris's contention that the atmosphere was created on the second day while planet earth was inserted into it on the third day (Morris, 1974, p. 208).

No witch doctor's potion could possibly comprise such illogical and contradictory ingredients as the elements outlined by Morris in his scientific version of Genesis. However, there are two of these which deserve particular attention because they are so widely used by creationists.

The first, and most obviously ridiculous, is that God dreamed up the law of conservation of energy while resting on the seventh day (Morris, 1974, p. 212).

The other assertion is that the second law of thermodynamics did not exist until fifty to a hundred years after creation. Morris states that this was done by God in response to Adam's disobedience. Before this, there was no pain, sin or death, but after it, God cursed the ground (Genesis 3:17) and took away from humans their immortality (Genesis 6:3).

Creation science uses a mangled version of thermodynamics to argue that evolution is impossible. The nonsense of their arguments is difficult for the nonscientist to detect since it is an unfamiliar area for most. However, a close look will show that Morris's manipulations of the universe don't come up to the high standards set by his Creator.

The Air We Breathe

People have sometimes wondered if a fish knows whether it lives in water. The same question in another form may be asked of us. Are we aware of the air we breathe? Are we aware of the gravity in which we are born, grow and have our being? For most of us the sometimes awesome power of a storm and, nowadays, the pictures of weightless astronauts as well as the blue atmosphere of the planet earth as seen from outer space are potent reminders of both air and gravity.

But there is an even more subtle and all-pervading law of nature which is with us every moment. It is a law that is as much taken for granted as the air we breathe or the force of gravity. This is a law of *common sense*, learned without being taught at a very early age. It is known by the intimidating title of *the second law of thermodynamics.*

If we watch a film of an exploding object, it is often amusing to run the film in reverse. We see fragments and pieces gather themselves together, fly through the air and reassemble themselves into the original object. Such a reversal from *disorder* to *order* contravenes our life experience. We consider it amusing because we know that what we are watching is impossible. Though we may seldom, if ever, think about the fact that we are watching a violation of the second law of thermodynamics, this is what is happening. We are watching disorder gather itself into order with no outside intervention. We know intuitively that this cannot happen in our universe. We learned from the first cereal bowl we smashed when a small child that disorder does not spontaneously reorganize itself into order. We have never seen an example of a smashed bowl reorganize *itself* so that the pieces formed a whole bowl again. This is not the way of our universe.

Unlike the fundamental constants of the universe, such as the speed of light, the laws of nature are statements obtained by observation of the behavior of nature itself. The most wide ranging of all the laws of our universe is the second law of thermodynamics and we observe its action almost every minute of our waking lives. It also defines our universe, but in a different way from the fundamental constants.

The second law does not have a mathematical formulation but simply defines the *direction* of change in our universe. Mathematics is, of course, used to examine further many of the implications of the second law.

The second law simply states that in any *spontaneous* change, the direction of the change is toward greater disorder. Thus, although we can imagine otherwise, we never observe a pile of bricks of their own accord *spontaneously* arranging themselves into a house.

Disorder is given the name *entropy*. An increase in disorder, such as the smashed cereal bowl, is referred to as an *increase in entropy* . On the other hand, for entropy to decrease, that is, for order to appear out of disorder, some input is required, usually in the form of energy of one kind or another. The pile of bricks which becomes a house required an input of energy from the builders. It is an example of entropy decrease. The house falling down is an example of entropy increase.

The second law may be understood in terms of probability. We would think it highly improbable that we could suffocate in a room full of air because the air in that particular room had spontaneously gathered itself into one corner of the room!

Although we might consider it possible for this to happen, unless there is some mental derangement we have learned to go confidently into strange rooms because we know that the likelihood of all the air clumping into one corner is so infinitesimally small as to be nonexistent. Such a change in the behavior of air particles would see the particles spontaneously shift themselves into a system of greater order. This contravenes the second law.

Does Life Contradict the Second Law?

On the other hand, by means of time-lapse photography we can see the growth of a plant from seedling through to an unfolding flower. This is an amazing sight of disorder organizing itself into order. Life appears to contradict the second law.

Although the second law states that spontaneous change is always in the direction of greater disorder, there is *one important condition*. This condition is that energy cannot enter or leave the object which is changing. In slightly more technical language, it is said that the second law applies only in a *closed system*.

The growing seed or tree is an example of an *open system*. Light and heat from the sun, moisture and nutrients from the earth, enable the living seed or tree to grow and become more complex, to fulfill the complexity inherent in the seed. As with ourselves, the cost, as it were, of holding the second law at bay is a constant input of energy in one form or another. Remove the energy input and decay, death and the second law take their course.

While we have the sun, the planet earth itself is an open system and life is possible. Take away the sun and life on earth would quickly cease.

Biological evolution is markedly different from stellar and chemical evolution because the advent of life signaled the existence of entities to which the second law does not strictly apply. Each living organism considered separately is an open system.

A further factor which distinguishes biological evolution is that living beings are users of entropy from their surroundings. They are islands of negative entropy requiring a constant energy input to maintain their high state of order or negative entropy. Most life-forms cannot use the energy from the sun to maintain order but must rely on those life-forms which can do this. The most notable of these latter are, of course, plants.

The energy which sustains life on earth has its source in the "burning" of hydrogen in our sun. As long as our sun has a surplus of hydrogen "fuel," life can continue to exist on our planet.

The Second Law and Time

While the second law forbids perpetual motion machines, it is also responsible for "time's arrow" in our universe.

The direction of time which we have is an interpretation of our experience with nature itself starting from early childhood. It is mostly from our sense of time that we "know" the second law of thermodynamics.

The "Heat Death" of the Universe

Much of the work which led to our present understanding of the second law was done in the latter part of the nineteenth century. It became clear from the second law that the universe would run out of heat and that our planet would be a frozen, lifeless rock orbiting in space.

This was sensationalized in the newspapers of France. They implied it could happen at any moment. Graphic drawings of huddled mothers with children, in the company of frozen corpses littering the lifeless and iced-over streets of Paris, caused a near panic in some sections of the population. Newspaper sales also increased.

What the French newspapers forgot to say is that this will not occur for a few billion years. It is this half-truth which seems to have stuck in the minds of creationists.

A living organism such as a tree or a human being exists in a highly improbable state of order compared to the rest of the universe. Energy is required for all the changes continually taking place as the organism constantly replaces all those parts which, in obedience to the second law, would otherwise decay and die. Without energy input the whole organism dies.

The law of conservation of energy also comes into play and the total amount of energy remains unchanged. One organism may use energy but its waste products and even the organism itself may provide food for other organisms.

However, in all the multiplicity of reactions and changes, heat is generated. Some of this is lost and eventually radiated into outer space where it is no longer available as "useful" energy. A person radiates heat at an average of about 150 watts. Six people together give heat roughly equivalent to that of a single-bar 1000-watt radiator.

The source of replenishment of heat energy for life on earth is, of course, the sun. When the sun runs out, "heat death" will ensue in our solar system. Heat death is defined succinctly as

> the proposition that the universe will come to an end when all its components are at the same temperature; the entropy of the universe will then be at a maximum and its available energy nil. This follows from the second law of thermodynamics if the universe is considered a closed system.

However, heat death as such will not occur until the sun uses up all its nuclear "fuel." We have quite a few billion years to go yet. Creationist publications and films indicate they haven't the foggiest understanding of this aspect of the second law.

What Use Is the Second Law?

Creationists view the second law simply as the cause of decay and death. They see it as God's punishment on creation because the first human beings disobeyed their Creator God. For the first fifty or so years after creation, they say, *the second law did not exist.* The world was perfect and there was no decay. One must admit that, at first glance, it seems reasonable to ask, why the second law?

Perhaps, for our purposes, the question may be best understood by considering for a moment the first law of thermodynamics. Creationists do believe that the first law existed from the beginning of creation, although Henry Morris (1974, p. 212) believes that the first law commenced on the seventh day, when God rested after finishing creation.

The first law of thermodynamics is also known as the law of conservation of energy. To give a common example: when a hot object and a cold object are placed in contact, heat passes from one to the other. The law of conservation of energy dictates that the amount of heat *lost* by the hot object equals the amount of heat *gained* by the cold object.

The first law is entirely sensible. No one has observed it being contradicted. It would be an obviously nonsensical world without the first law. Even creationists understand the first law and admit that it existed from the beginning of creation, although they state that the second law began later.

Let's consider our hypothetical hot and cold objects. The law of conservation of energy is also obeyed if the cold object becomes colder and the hot object becomes hotter. This, of course, does not happen in our experience. The question is, if the law of conservation of energy does not explain why the hot and cold objects arrive at the same final temperature: what does explain it? The answer, as you may have guessed, is the second law.

A red-hot nail is at a higher *temperature* than a tub of boiling water. Both contain heat energy. But the *heat* energy of the water in the tub is much greater than that of the nail. Putting it another way, the nail will become red hot (600°C, 1112°F) in seconds when placed in the flame from a gas stove. The tub of water may take an hour to boil (100°C, 212°F) using the same flame. The *quantity* of heat in the tub is greater than in the hot nail but the *degree* of heat is greater in the nail. Thus it can be said that an iceberg contains more heat than a bucket of molten steel.

When substances are heated they become more disordered. Ordered crystals of ice gain heat energy to become less ordered molecules of liquid water. With more heat, they change to fast-moving single molecules of gaseous water, or steam. The *heat energy input* increases the entropy or disorder as the *temperature* increases. But the amount of disorder (or entropy) is not only a function of the amount of heat *but also of the temperature*. The important factor is that heat causes more disorder (or entropy) when added to cool objects than when added to hot objects.

Suppose that hot water is added to cold water. Although the hot water gives up heat (and hence becomes more ordered), the heat gained by the cold water, because it is at a lower temperature, causes more disorder than that lost by the hot water. When the final temperature is reached there is a net gain in the amount of disorder, and thus maximum disorder for the two objects taken together is achieved. The two objects in this case are the hot water and the cold water. Therefore, the direction of heat flow is a less obvious consequence of the second law.

To summarize, the flow of heat from hot to cold occurs because in this way the disorder of the system is increased. Maximum disorder is reached when both objects arrive at the same temperature. The *energy* of the system remains the same but there is a net gain in *entropy* (or disorder). This is a spontaneous change in accordance with the second law.

What creation science fails to understand is that without the second law, life itself would be impossible. The most obvious argument from thermodynamics is that the second law indicates the direction of flow of energy. Without the second law a cold object and a hot object in contact *might* arrive at the same temperature. But, on the other hand, the hot object *could* become hotter and the cold object colder without contravening the law of conservation of energy (the first law of thermodynamics).

For creationists to state that the first law existed at the beginning of creation, and the second law fifty or so years later, is a total contradiction. You can't have one without the other.

Such an ignorance of basic physics precludes creation science from any claim to be "science." It marks creation science indelibly and forever as a *pseudo-science* . It is nigh impossible to deny that some creation scientists at least are propagating this error knowingly. They have been doing it for well over a decade. Elementary thermodynamics is not advanced physics but is covered in some high school physics courses and in all first-year university physics courses.

A universe without the second law of thermodynamics, which creationists claim was the situation for the first fifty-odd years after creation, is utterly nonsensical – something Alice would encounter in Wonderland.

The basic error of creation scientists in their misunderstanding of thermodynamics is, by itself, sufficient to demolish any claim for creation science to be called science. This "universe" of creation science is a fantasyland where Alice and the Queen can remember tomorrow but not yesterday!

Chapter 11

Some Current Issues in Science

Creationists are quick to point to any debate or disagreement in any area of science as evidence that scientists don't know what they are doing or that such debate indicates that evolution, the age of the earth, radiometric dating, the fossil sequence, and so on, are shaken to their foundations. Science thrives on continued unearthing of new knowledge which as often as not leads to what may not be fully understood for many years. Basic research leads to further research. The scientific enterprise is always in an unfinished state. This is both the essence of and the fascination of science. The creationists' world of science is one which is no bigger than their own minds. The seemingly contradictory particle and wave theories of light existed side by side for many decades. Paradox is a sign of truth.

We have long divided the universe into living and nonliving. We have said, and believed, that only life can beget life. But scientists are becoming increasingly aware of the inherent ability of matter and energy to self-organize.

Far Cry From Darwin

Biological evolution is in a very healthy state. The understanding of it is increasing day by day. Molecular biology and genetics, the discovery of more and more transitional forms in the fossil record, the finding of more and more fossils, including those in the oldest of rocks, are currently some of the areas which are increasing the understanding of the mechanism by which life has evolved. The concept of punctuated equilibrium – that the evolution of life occurred not gradually but in quick changes followed by long periods of stability – is gaining ground.

There is not one new finding which casts doubt on the fact that evolution has occurred. The debate about *how* it has occurred is the very stuff of which science is made. It is quite freely admitted that there are factors operating in biological evolution which are presently unknown or barely hinted at.

Evolution is the conceptual framework into which is set all the hundreds of branches of science. There is no shred of evidence for any theory of origins but evolution. Nor, given the evidence available today, is it possible to imagine any other theory which could account for the mountains of evidence from every branch of science which make sense *only in the context of evolution.*

However, there are two major areas of debate at this time which seem likely to shed light not only on the mechanism of evolution across its whole spectrum of 15 billion years but also on the nature of the cosmos itself. These are holism and the concept of "order out of disorder."

Order Out of Disorder

One area of particular interest concerns the research of Ilya Prigogine, a Belgian scientist and winner of the 1977 Nobel Prize in chemistry. He describes this in his book *Order Out of Chaos*, co-authored with Isabelle Stengers and with a foreword by Alvin Toffler. Prigogine is concerned with the thermodynamics of systems which are far from equilibrium, that is, in a high state of disorder. The second law, which creationists mistreat with such gay abandon, is strictly applicable only to closed systems. These are systems or portions of the universe which are closed off so that energy cannot enter or leave. Planet earth is not a closed system as long as it continues to receive energy from the sun (see Chapter 10).

Prigogine's work indicates that under appropriate conditions, disorder may spontaneously produce order, contrary to the conventional understanding of the second law of thermodynamics. It is apparent that the findings of Prigogine may be of the utmost importance in furthering an understanding of how the universe has evolved from the simple to the complex. His results indicate that the production of order from disorder occurs when disorder is greatest, when the system is far from equilibrium. It is under these conditions that order, organization and life itself may be initiated.

Prigogine's work has enormously wide inplications for all areas of science and, it is claimed, for society as well. If Prigogine is right then both chance and determinism are playing a part in the infinite number of dynamic processes taking place at every moment in the universe.

One can do no more here than skim the surface of this dramatic new addition to the understanding of thermodynamics. In his conclusion Prigogine states:

Until recently, however, there was a striking contrast. The external universe appeared to be an automaton following deterministic causal laws, in contrast with the spontaneous activity and irreversibility we experience. The two worlds are now drawing closer together. Is this a loss for the natural sciences? (Prigogine, 1984, p. 31)

What part and to what extent Prigogine's findings play in elucidating the seemingly unknown factors operating in evolution, only time will tell.

Physics and Holism

One of a number of factors which distinguish today's physics from that of the past is the change from reductionism to holism. *Reductionism* is the view that by pulling an object apart and understanding the bits and pieces, you then understand the whole object. Newtonian physics was reductionist. It led to the view of the universe as a predictable machine, a clockwork universe. Holism as used in modern physics asserts that the whole is greater than the sum of the parts: two plus two add up to five, even six or seven or eight. The atom is a perfect example of this.

The simplest atom is hydrogen, which has a single proton and a single electron. The proton is a particle with a positive charge and a mass 2000 times greater than the electron, which has a negative charge. The hydrogen atom consists of a proton as nucleus and one orbiting electron at a relatively vast distance from the nucleus. The hydrogen atom is neutral because electrons and protons have equal but opposite charges. Most of the volume of the atom is empty space between the positive nucleus and the orbiting electron. If the space were removed, this book which you are reading would be smaller than a speck of dust but its mass would be the same. Protons themselves are made of other particles but this need not concern us here. The neutron is a neutral particle made up of a proton and an electron.

Yet if the proton and the electron are considered as separate entities, it is not possible to predict that these two specks of matter could form a hydrogen atom. The atom is much more than a proton plus an electron. It has properties of its own which far exceed the combined properties of its parts.

If we have a nucleus of two protons together with two neutrons and a pair of electrons orbiting around them, the resulting substance is not like two protons and two electrons; nor is it like hydrogen. It is an atom of the inert gas helium, commonly used in airships because it does not burn or explode. Taking it further, three protons, three electrons and the appropriate number of neutrons (four) make an atom of lithium, a soft silvery-colored metal which explodes on contact with water.

Moreover, when photographed by modern techniques single atoms appear as a blur. This is the electric field surrounding the oscillating nucleus. One would be hard put to predict that en masse the properties of color, taste, solidity, and so on, could emerge. Conversely, it is difficult to imagine the ink and paper of this page as points of matter unceasingly vibrating. The 92 elements found in nature are from the same building blocks. They can be transmuted one into the other. The *arrangement* of the same few components gives unique properties to each.

But the above is like a tune played on one note compared to what happens when different atoms combine with one another. The new properties are equally unpredictable. Metallic, silvery sodium atoms react violently with the atoms of poisonous green chlorine gas to form the white crystals of sodium chloride, common salt. Gaseous hydrogen explodes with colorless oxygen to form hydrogen oxide, water. Water is a substance which should by all rights be a gas at normal temperatures. And it would be except for the unique quality of hydrogen which enables it to form weak bonds with other water molecules. Without this "hydrogen bond" there could be no life. Yet the constituent parts of a molecule of water are 18 protons, 18 electrons and 16 neutrons.

Throughout the world of atoms the whole is greater than the sum of the parts. It is not just that the whole exceeds the sum of the parts but that examining the components provides at best a partial explanation. In the case of even relatively complex molecules the existing theories of chemical bonding can both explain and predict molecular properties to a large extent. A stage is reached when it seems that a given arrangement will result in molecules which can reproduce their own kind. These self-replicating molecules give rise eventually to that mysterious quality we call life, the most unpredictable of the properties which seem to be potentially present in matter.

This is not to say that "life" is simply an arrangement of matter, even for those primitive single-celled life-forms which first appeared 3 million years ago. It is saying rather that under certain conditions and arrangements, a new property appears which is as unexpected as any other of the new properties mentioned.

The above is an extremely simplified but hopefully not oversimplified attempt to indicate one way in which physics today has changed from the mechanistic and reductionist approaches prevalent in the past.

Paul Davies uses the analogy of a newspaper picture of a face composed of a myriad of tiny dots to illustrate the concept of holism which is part and parcel of physics today:

No amount of scrutiny of the individual dots will reveal a face. Only by standing back and viewing the collection of dots as a whole, on a coarser scale, does the image emerge. The face is not a property of the dots *themselves*, but of the *collection* together – it is to be found in the pattern, not in the constituents. Likewise, the secret of life will not be found among the atoms themselves, but in the pattern of their association – the way they are put together, in the information encoded within the molecular structures. Once the existence of collective phenomena is appreciated, the need for a life-force is removed. Atoms do not need to be "animated" to yield life, they simply have to be arranged in the appropriate complex way. (Davies, 1983, p. 61)

A holistic view of life is in direct opposition to a reductionist view which sees life as nothing but a collection of atoms and molecules. Neither does it require the invention of a "vital force" or the introduction of the "god of the gaps" as used by creationists and some theologians.

Probabilities, Possibilities and Impossibilities

While interviewing Duane Gish after he had notched up a hundred straight wins in debates with academics a reporter asked him the reasons for this remarkable success. He replied: "I go for the jugular."

From Fish to Gish, by Marvin L. Lubenow, is a creationist account of Gish's debates and strategies. In a recent review of this book Russel Grigg (1988, p. 19) lists the topics on which are based the strategies used by Gish and Henry Morris, whether debating together or alone:

• The laws of thermodynamics show that evolution *could not* take place.

• The fossil record shows that evolution *has not* taken place.

There is another strategy used by Gish but not mentioned in Grigg's review. Gish invariably uses an argument from probability. His favorite is to cite an animal that changes form dramatically during its life cycle. For example a butterfly. The cycle begins with the larva (caterpillar), then to pupa in a cocoon, from which emerges the adult (butterfly). Without a doubt this is a wondrous and impressive example of the diversity of life.

Gish follows his exposition by asking how can this be explained by genetic mutations over millions of years when most mutations are lethal; that is, the resultant mutant dies. Gish states that it is patently apparent that a considerable number of simultaneous mutations would be required. He then estimates the probability of so many genes mutating simultaneously and arrives at one chance in 10 to the power of 40, which is 1 followed by 40 zeros, a number so improbable as to border on impossibility.

Therefore Gish concludes that the butterfly could not have evolved but must have been created as it is. This is a difficult argument to rebut because of the hidden assumption that simple mathematics can be used willy-nilly on complex living organisms. Life is full of examples similar to the butterfly. Creationists argue similarly that the evolution of the eye demands too many simultaneous mutations to be explained by evolution.

What Gish is really doing is stating a reason why randomness, natural selection and genetic mutation are not *sufficient* to explain completely the mechanism of evolution in such cases. Biologists would nowadays mostly agree with this. However, they would also assert that there are other, not yet fully understood, factors operating in evolution. Gish will need to find another example because, as mentioned in Chapter 4,

> The metamorphosis of butterflies is becoming quite well understood in terms of molecular biology. There are specific hormones, such as ecdysone and the juvenile hormones, that have been chemically synthesized, and these turn genes on and off, enabling metamorphosis to take place. (Jukes, 1988)

The sophistry in Gish's statement is in his applying the statistical arguments of mathematical probability to a situation entirely unsuited to such arguments. His reasoning supports the opposite of his conclusion. The butterfly does exist in spite of Gish's arguments. This is, if anything, evidence that there are other factors operating which render invalid the simplistic application of mathematical improbability. If butterflies did not exist Gish *might* have an argument.

This leads us to consider the physical constants which seem to define our universe as unique. "Fundamental constants are certain constants that play a basic role in physics and which have the same numerical value everywhere in the universe and at all moments in time" (Davies, 1983, p. 187). Examples of universal constants are an atom of hydrogen, the speed of light and the gravitational force. Why these constants have the values they do is totally mysterious. They are of crucial significance for the structure of the physical world. But what is also amazing is that a small change in value of any one of these constants shakes the universe to its very foundations.

The "strong nuclear force" is also a universal constant. It acts to hold together the nucleons, or particles which make up the nucleus of an atom. It acts over a very short range. Without it there could be no atoms other than hydrogen, which has a nucleus of a single proton. There could be no universe.

After hydrogen, deuterium is the next simplest atom. Deuterium has all the chemical properties of hydrogen because it has, like hydrogen, a single electron in orbit around its nucleus. The only difference is that it has a neutron as well as a proton in its nucleus. It is twice as heavy as the "normal" hydrogen atom. A small proportion of seawater is deuterium oxide, whereas the rest is hydrogen oxide. Deuterium oxide has been given the name "heavy water" and is important for the functioning of one type of nuclear reactor.

Heavy water is the moderator used in HIFAR, Australia's research reactor at Lucas Heights near Sydney. A moderator slows down neutrons so that fission can occur. The advantage of heavy water as a moderator is its safety. If a leakage occurs and the heavy water is lost, the reactor simply shuts off. Without a moderator there is no reaction.

The single proton and the single neutron in a deuterium nucleus are held together by the strong nuclear force, but only tenuously. If the value of the strong nuclear force was only a few percent weaker deuterium could not exist. Since deuterium is a link in the nuclear reactions of the stars they could not exist. There would be no sun, no life. Moreover, if the strong nuclear force was a few percent stronger, there would be no hydrogen left over from the big bang and neither stable stars like the sun nor even liquid water would exist. The precise value of the strong nuclear force, like the value of other constants, is critical to the existence of the universe.

Another example is the value of the gravitational constant. If it changed either way by as little as one part in 10 to the power of 40 (1 followed by 40 zeros), stable suns like ours could not exist. The list of such examples is very long. Once might be an accident but not so many (Davies, 1983, p. 188).

Gish argues from an improbability factor of 10 to the power of 40 against the butterfly having evolved to conclude that evolution is impossible. If we apply Gish's line of argument to the physical constants of the universe instead of the butterfly, it would follow that the universe cannot exist. Thus Gish himself could not exist.

If we accept the validity of arguments from mathematical probability, neither the butterfly nor the universe as we know it can exist. But they do exist. Where is the flaw in the argument? As shown for the butterfly, the flaw is that there are factors over and above randomness operating in the evolutionary processes which apply to life. Biologists believe this.

In the case of the values of the universal constants alone, there is a seemingly endless list of such improbabilities. When these are combined, it becomes impossible to hold that the universe is a result of chance. It does not follow, as Gish would argue, that evolution is therefore false and that the world was made as described in Genesis. What is concluded is that matter itself possesses the potential for life. Matter is self-organizing.

Life and Planet Earth

One can go further and wider than a consideration of the universal constants in isolation. It is apparent that if those same universal constants had been different at the beginning there would be no universe. But there are many other factors involved besides the constants.

If the primal explosion had been smaller or larger there would have been no universe. Likewise if the temperature after the big bang had been hotter or colder. Or if the swirl of matter that condensed to our solar system had taken a different arrangement; or if the sun had been larger or the moon smaller; or if the sun, moon and earth distances had been different; if continental drift had not occurred in the way it did and at the pace it did to make our oceans; if the earth was not inclined to its orbit; if early life-forms and the primal atmosphere had not interacted the way they did; if the ozone layer had not formed, or the Van Allen belts ... Or if ..., or if ... for hundreds of known phenomena and probably thousands of unknown.

The conclusion is not only that the probability of it happening by random choice is so small as to be impossible. Nor is it that our universe had within itself from the very beginning its own plan. Nor is it that the chances of life on earth are so improbable, and the 15 billion years to get it right so improbable, but that it could not happen again.

The major conclusion which is highly relevant to our present discussion is that, as far as it is possible to know, *it couldn't have happened any other way* .

One can invoke the possibility of an endless number of universes with our universe being the one that got it right. One can validly conclude from the physics of the universe the presence of design or meaning. But this is as far as science can go. To conclude further that there is a god involved is to step across the boundary between science and religion. It becomes a matter of faith, which is an individual decision.

Stepping Across the Boundary

Many scientists through their science find for themselves faith in the existence of a god. Many do not. Einstein states his own beliefs thus:

> I believe in Spinoza's God, who reveals himself in the harmony of all things, and not in a God who is interested in the actions and the destiny of each individual ... The ideals which have lighted my path at every moment and which have given me the courage to look at life joyously have been beauty, goodness and truth. Without brotherhood to unite

me with men having the same ideal, without my interest in the structure of the world and of what is inaccessible in the field of art and science, life would have seemed empty to me. (Chagas, 1979, pp. 73, 74)

However, it is the intention of this section to not merely step across the boundary but to go back in time to a different mindset about creation and the making of the world. This will disclose a core belief in creation science, to be defended at any cost.

In the early centuries of Christianity, given the limited cosmology of the time, it is easy to see how it would be accepted without question that an all-powerful god could have created any number of different worlds. In addition, assuming the same god was infinitely good and wise, then this world must have been the best of all possible worlds.

The Eastern mindset is far more cautious than that of the West when it comes to making conclusions about such deep metaphysical matters. The *Rig Veda*, one of the great Hindu sacred books, also discusses the same question of the origin of the world, as do all religions in one way or another. The author of the *Rig Veda* refuses to commit himself to a definite answer, preferring to let alone such a deep mystery:

Whence all creation had its origin,
he, whether he fashioned it or whether he did not,
he, who surveys it all from highest heaven,
he knows – or maybe even he does not know.
(Quoted in Eliade, 1967, p. 130)

Belief in a duality of good and evil was very common in ancient times in the Middle East. The author of the first chapter of Genesis takes a firm stand against the belief in both good and evil gods prevalent in the religions of the nations who were Israel's neighbors. At the end of each of the six days of creation he quotes God as saying that He was pleased with what He saw (Genesis 1:4, 10, 12, 18, 21, 25, 31). Thus he proclaimed unflinchingly that the God of Abraham and Moses was the author of all creation. Not that this was without problems.

About the same time that Genesis was written the author of the Book of Job (c.600 to 450 B.C.) discussed the problem of evil and a good god in what is still one of the masterpieces of world literature.

The story of Job is written at two levels. The reader is privy to the thoughts of God, who boasts at a meeting of the heavenly court about His servant Job:

"There is no one on earth as faithful and good as he is. He worships me and is careful not to do anything evil." (Job 1:8)

Satan, who is on conversational terms with God and also present at the court, challenges God that Job's faithfulness is simply because of what he gets out of it. Satan claims that God has showered Job with riches, blessings and enough cattle to fill the whole country. God rises to the bait and permits Satan, who is purely a literary device in the story and a far cry from what he was to become in Christian times, to test Job but not to hurt him. Satan claims he will make Job curse God to His face.

Job subsequently suffers an horrendous series of afflictions. All his property is destroyed, his children killed. He is covered with sores and cannot keep down any food; he is eventually driven out of town to live on a rubbish heap. He is totally unaware of the heavenly byplay between God and Satan. He curses the day he was born but staunchly refuses to curse God.

A main theme in the story is that three of Job's friends call to see him. They tell him that his predicament is because he or his ancestors have sinned in some way against God. They urge him to ask God for forgiveness. But Job proclaims his innocence and refuses to ask forgiveness for something he has not done. The dialogues between Job and his friends cover all the traditional reasons for evil in the world. Job demands to know from God why it is that evil men prosper and the good suffer. He adamantly refuses to accept that God has treated him justly but still refuses to curse God.

Eventually God appears in the form of a storm to answer Job. He reminds Job of His power and majesty, and commands Job to

"Stand up now like a man,
and answer my questions.
Are you trying to prove
 that I am unjust
to put me in the wrong and
and yourself in the right?
                    (Job 40:7, 8)

Job admits his foolishness. God is angry with Job's friends for not speaking the truth and orders *them* to make a sacrifice. God then reinstates Job to his former position. *But God never answers Job's question as to why the good suffer and the evil prosper.*

To expound upon the richness and depth of meaning in the Book of Job is beyond any single book, let alone a short summary. The purpose here is to make a few points relevant to seeking a further understanding of what it is that underpins creationism. After forty-two chapters the author of Job comes not to an answer but to a conclusion: that the ultimate causes of good and evil are a mystery known only to God a mystery too deep for the human mind to fathom.

Another point is also apparent. There is no suggestion whatsoever that evil came into a perfect world when Adam and Eve sinned in the Garden of Eden. This supports modern biblical scholarship that Genesis is not meant to be historical, nor is it to be interpreted literally. It is to be read in other ways, some of which have been mentioned previously.

A thousand years later, the mind of the Christian West accepted Genesis as literal truth. Combined with the mindset that this creation is the best of all possible worlds, conclusions were reached and doctrines forged which were to have a major effect on Christianity for another thousand years and even up to the present. This will be discussed in Chapter 15.

Chapter 12

Creation Science
in the Classroom – A Case Study

This chapter considers the effect of creation science in the classroom. A case that became a nationwide scandal. Many of the student materials have not been published before. The details of what happens in actual classrooms where creation science is taught are nigh impossible to obtain. What happened at Livermore, California is a damning indictment against creation science. Creationists know this and fear it more than all the scientific and religious arguments put together.

It is an object lesson for parents everywhere. It should cause those parents who have sent their children to Christian schools, in order to learn Christian values, to check closely that what happened at Livermore is not happening to their children. *In some cases it is.*

All parents should be concerned whenever creation science, in any form whatsoever, is allowed into schools. If parents are unaware of what could happen, then surely there will be concern after reading this.

The Emma C. Smith School

In 1980 the Emma C. Smith Elementary School at Livermore became the center of a storm of controversy which sent shock waves throughout the State of California and across the nation. A group of sixth grade students had been taught a course in creation/evolution by Ray Baird, using materials almost exclusively from the Institute for Creation Research: texts, slides, videotapes and newsletters. Baird had been teaching the same course to classes of 12-year-olds for two years prior to this.

The materials had been bought with state funds intended for gifted students. It was alleged by one student's parent, Marian Finger, that Lloyd Teel, then principal of the school, had signed the requisition for the materials, ignoring a complaint made by one of the teaching staff about the religious content in the course called "Creationism - Evolutionism."

In the United States it is illegal to teach a particular religion in public schools, although schools may teach about different religions in a historical context. The separation of church and state is a strongly upheld tenet of the Constitution (Amendment I, Bill of Rights). Many of those who first settled in America had fled from the religious wars and persecutions of seventeenth century Europe.

The Livermore Parents

There are hundreds of newspaper clippings on every aspect of the Livermore affair. The following indicates the feelings of some of the parents:

> The situation in Livermore developed slowly over a period of three years, according to Finger. During that time, Finger says, a sixth grade teacher, Ray Baird, ordered about $700 of materials, primarily from the Institute for Creation Research and the Creation Science Research Center, and was using the material to teach a unit on creation/evolution. The money came from the state grants to California school districts to be used for materials for gifted children.
> Finger's son Eric took the unit in 1979. On the first page of Eric's notes for the class is a statement of the Genesis account of creation. According to Eric, his teacher emphasized the mutual exclusiveness of the two models. "He said that either both were wrong, or one, but not both, could be right," Eric says.

Eric says that at the end of the unit, the teacher conducted an anonymous vote in which the students had to choose between evolution and creationism. According to Eric, six students out of a class of about 30 voted for evolution; the rest voted for creationism. The teacher presented a tally of the votes to the class.

"Most of those for evolution were among the gifted students," Eric says, who is himself among the gifted students. "I thought a lot of the others were maybe influenced by how Mr. Baird presented it. I don't know if that's true, but it could be."

Eric and the rest of the gifted students were given extra assignments in the class: They were required to view filmstrips over again in the library.

Those filmstrips were blatantly religious, according to Sheila Karlson, another of the mothers. "One of them started out by saying, and this is almost a direct quotation, that 'Either the Bible is true or evolution is true. You must make a decision.' It goes on from there to give this very distorted picture of evolution and this glowing picture of creation."

Karlson became aware of the situation when her daughter, Kristine, became worried about preparing a report in which she had to present the scientific evidence for both evolution and scientific creation. "Scientific creation was a term I had not heard of," Sheila Karlson says. "I didn't know where to get that information. I told Kristine to go to the Bible, but she said, 'No, it has to be scientific evidence.'"

Kristine Karlson was convinced that creationism was correct. "Lots of people believed in creationism," she says, "because the teacher is such a nice, friendly person."

"I saw that as the biggest threat," Sheila Karlson says. "Here you have a teacher who is worshipped by his students. He could tell them anything and they would believe him."

Karlson says that she is uncomfortable with forcing children to make such an important choice. "I was trying to make clear to Kristine that I don't know how many people ever make that choice, and that it isn't necessary."

When Sheila Karlson approached the teacher about the

course, another mother went with her because, Karlson says, the woman's son refused to watch the television program "Cosmos" because, her son said, Carl Sagan was a liar, according to the teacher of the class.
Ruth Anne Hunt's son John Patrick is a bright, articulate, 12 year old. John Patrick described what his teacher said prior to showing a videotape of a "Cosmos" presentation on the evolution of crabs. "He poisoned the well," John Patrick says. "He would say: 'Television people manipulate you' ..." (Baum, 1982, p. 24)

Another newspaper reports Baird as saying that the resources used to teach his course were just placed on the shelves and not used in class presentations. What he omits to say is that the shelves were in the library. The books were borrowed extensively by the students for reports and assignments. There was nothing else they could use. There was considerable divergence between Baird and parents on this issue.

Baird has admitted to error in using these materials, but has contended that the course had already started when they were delivered and he "just placed them on the shelves without really reviewing them," a failing he has also called an error. He has said he incorporated none of these materials in his class presentations, and he gave "equal, if not more, time and consideration" to the theory of evolution.
The protesting parents give a different picture. They say Baird's course was slanted to creationism and his presentation of the issue conformed to the ICR's teachings, in particular its doctrine that the choice between creationism and evolution is a choice between God and atheism. The parents also contend that Baird was fully aware of the content of the materials, some of which had been in his classroom since last year. (*The Independent,* 7 January 1981)

The point overlooked by both Baird and parents is that these materials or their equivalents from the ICR are the only ones he could have used. They are designed and published for schools and without them there would have been no course. He would have been doing no more than standing up in the front of the class voicing his own opinion.

There is more than a little doubt that Baird gave equal time to evolution and creationism.

> "I think it's true he gave more time to evolution, says one parent. "He spent 40% of the time telling the kids why creationism is good and the other 60% telling them why evolution is bad."
> Another parent, whose child observed Baird teaching this subject three years ago, relates that while Baird succeeded in winning some converts to the creationist view, other students, including her own child, were so appalled that they completely rejected religion in their own lives. According to this mother, all the teacher really accomplished was to polarize the class into two camps, the believers and the nonbelievers. (The *Independent*, 7 January 1981)

One of the mothers writes:

> The most dangerous information to the scientific creationists was the fact that the gifted students could see how bad the science was and that they were voting evolutionism which was, in the context of the course, the same as voting atheism. Some of the gifted students voted evolutionism because they could see the fallacy of the either-or approach. Some actually, in anger, did give up religious belief. (Finger, 1988)

The Materials Used

Between 6 March 1979 and 22 January 1981, invoices from Creation-Life Publishers, San Diego, addressed to Livermore Valley Unified School District (California), showed a total of $735.29. Of this, $345.95 was spent on the purchase of eight filmstrips and a set of transparencies. There was also a purchase of *Science Readers* from the Bible Science Association. These are in sets of eight, one for each week of the course. These are all marked Public School Edition.

In creation science publications the only difference between Christian school and public school editions is the erasure of biblical references. With the exception that Christian school editions may sometimes have short religious messages at the end, both editions are identical. Not all creation science books come in two editions.

The majority of the materials ordered were obviously not intended for classroom use. They included technical monographs and other publications with titles such as *Genetic Engineering* and *Ageing and Freezing Human Bodies*. Most of these seem intended for personal use as background material for Ray Baird, who admits he is a convinced creationist. He is also, by all reports, a very popular and effective teacher.

The only books ordered in multiple copies were *Dinosaurs, Those Terrible Lizards* (10), *Evolution? The Fossils Say No (Public School)*, (4) and *Dry Bones ... and Other Fossils* (7). The first two are by Duane Gish and the third is by Gary Parker. There were also single copies of books by Henry Morris, for example *The Twilight of Evolution*. This book is especially renowned for Morris's derivation from the Bible that the craters on the moon are scars from a primordial battle between two factions of angels, and that flying saucers are in reality the devil's minions visiting earth from their homes in the asteroid belt. The most popular books with the students were those by Gish and Parker.

There is not a single book in the eight weeks' course which presents the viewpoint of evolution. Not one! Even though Baird claimed he was "only showing kids how to make choices between two contrasting viewpoints" (Livermore *Herald*, 18 January 1981).

In the Classroom

The children's notebooks plus the texts used for the course are graphic. They tell very clearly the teaching objectives and strategies for the unit, and they illustrate the pseudo-objectivity of the teacher:

- There are two models which are used by science and each is valid.
- Neither model can be proved. Each is a matter of faith.
- One model is followed by those who believe in God. The other model is followed by atheists.

The materials, both textual and visual, the constant repetition, the conviction of the teacher, the absence of any other viewpoint all add up to little short of brainwashing when drummed in over a period of eight weeks by a competent and trusted teacher with a class of 12-year-olds.

It is important to examine a sample of the work done by the students. Here are notes written by one student in the first lesson of the course:

Creationism
Creationists believe, in the beginning a creator created the heavens & the earth & all life in this order:

Day 1 The heavens & earth were created. All was dark at first, & then there was light for daytime. There was some form of water.
Day 2 Water divided from heaven & earth.
Day 3 Water separated from dry land to form oceans & land. Vegetation on land.
Day 4 Two big lights were formed – one for day & one for night.
Day 5 Sea life & birds were made.
Day 6 Animals arrived – males & females.
Day 7 The creator rested.

It happened in a short period.
Plants & animals were created in groups called kinds. There could be changes within a kind, but one kind could not change into another kind.
They also believed at one time there was a world wide flood. Before the flood there were no mountains & it never rained, there were only rivers to drink from. There was a warm even temperature all over the earth.

What caused the flood was there was a water vapor canopy all around the earth & it would protect all life from harmful rays of the sun.

Rain drops need a nucleus or a fine spec of dust or something solid. Water vapor is not clouds, steam or mist; it is clear water. If it ever started raining the whole water vapor canopy would come down on the earth.

At the time there were many volcanoes all around the earth. A volcano does not have to be tall. At the time a lot of volcanic action had been going on; a volcano erupted & somehow blew specs of dust into the water vapor canopy & it started to rain & the whole water vapor canopy came down & flooded the earth.

THE END.

And this was the teacher's comment on the notes:

Where is the scientific evidence that you have researched to support the theory of creationism. You have generally explained the theory well.

It is quite apparent that God has not been mentioned to this student. Evidently this is done so that it can be stated that it is not religion! In this way, Creator is made into creator – offending to many Christians. It is blatantly obvious that this is the biblical account of creation followed by a pseudo-scientific story of the Flood. It is found only in creationist writings and nowhere else.

This is what creation science calls an alternative account to evolution; it is claimed to be "not religion."

The following is an excerpt from the book most popular with the 12-year-olds, *Dinosaurs, Those Terrible Lizards:*

There are other scientists, called creationists, who believe that the scientific evidence shows that dinosaurs did not evolve, but that they were created by God, just as described in the Bible. Creationists believe that dinosaurs were created the same time that Man and all other creatures were created, probably sometime less than 10,000 years ago.

The Bible says:

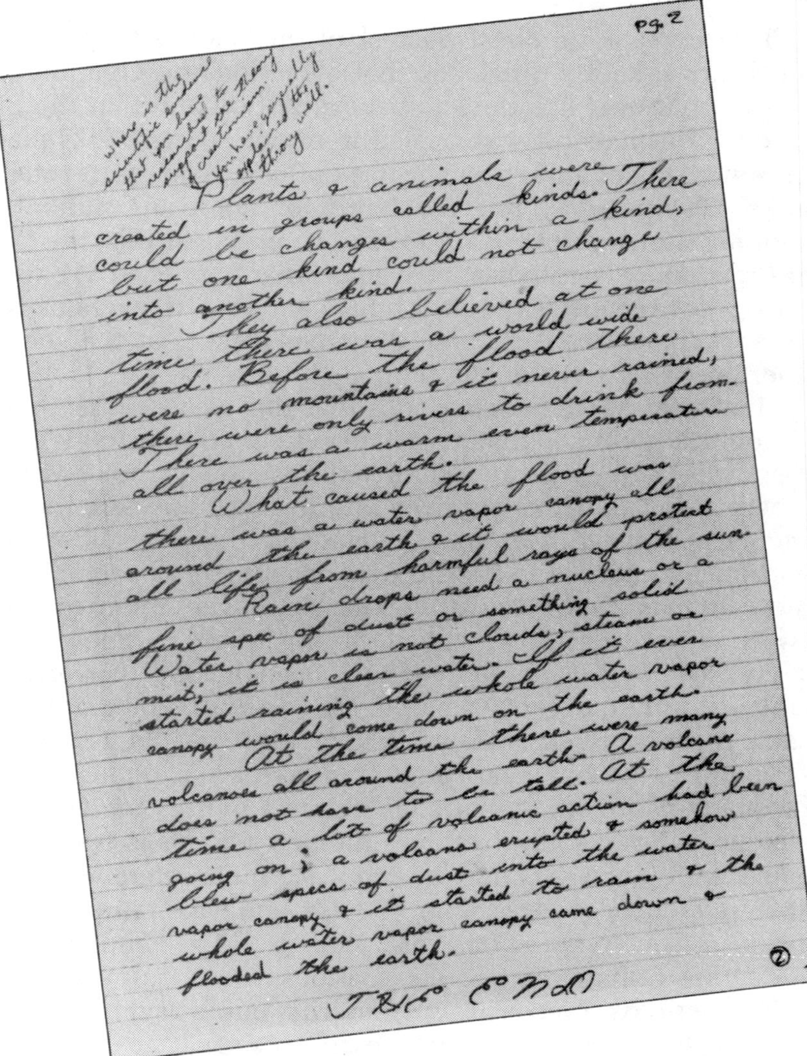

And God said, Let the earth bring forth the living creature after his kind, cattle, and creeping thing, and beast of the earth after his kind: and it was so.
And God made the beast of the earth after his kind, and cattle after their kind, and everything that creepeth upon the earth after his kind: and God saw that it was good. *Genesis 1:24, 25* . (Gish, 1977, p. 13)

This excerpt, as does the rest of the book with its appealing illustrations, subtly proclaims that evolution and creation science are on a par, that creationists believe in God and that the Bible is science. This excerpt was copied from one of the books placed in the Smith school library for the use of Baird's class. After the storm broke, Gish publicly rejected Baird for not using public school editions. He used this as an excuse to proclaim that the Institute for Creation Research disapproved of what happened at Livermore. It was a convenient way to wash his hands since this book by Gish seems to have been the only Christian version at the school. All other books were the public school editions.

However, there is some doubt that this was a Christian school edition. ICR policy is that Christian school editions may use Bible references but public school editions may not. Many ICR publications appear only in the one format. However, there is no evidence, at least in existing catalogues, that the above book of Gish's has appeared in both public and Christian school editions. The system is somewhat flexible because at the Arkansas hearing a few months after the incident at Livermore a copy of Gish's *Evolution? The Fossils Say No* had been hastily converted from public to Christian by the simple expedient of gluing a disclaimer inside the front cover (Gilkey, 1985, p. 301).

Following Gish's logic, what is nonsense in public schools is transmuted to truth in Christian schools by the *addition* of Bible references. Conversely, truth in Christian schools is transmuted to nonsense in public schools by *removing* Bible references. Hence a Bible reference changes true to false and vice versa. The mind boggles. The possibilities are endless. The implications for students in Christian schools are serious.

The following excerpt is from the public school edition of Parker's *Dry Bones ... and Other Fossils* (p. 68):

You didn't believe in God, Dad?
Not really, Dave. I surely didn't believe that God was speaking to me through the Bible. And, after all, there is one big weak spot in the creation view.
What is that?
You can't believe in creation if you don't believe in a Creator. If there is no Creator, then creation doesn't make any sense.

But if there is a Creator, then evolution doesn't make any sense.
You are right about that for sure, David! Evolution is not based on the fossil evidence. It is really based on belief that you have to explain everything without God.
So, creation and evolution are both really a matter of faith.
When you get to the bottom of it, that's true, Dave. If you don't believe in God, then you have to believe in some kind of evolution. And then you must try to make the fossils fit in with evolution somehow.
But if you believe the Bible, then you can see how easily the fossils fit with Creation, man's sin, and the Flood. Right, Dad?
Right, Dave!

The above excerpt shows without the shadow of a doubt that not only are young children being forced to make a choice between God and evolution but they are being taught that the entire worldwide community of scientists is totally deluded because they don't believe in God. These palpable lies are repeated ad nauseum through the seventy pages of the book. The actors in the charade are Dad and his son, Dave. Dad is a former atheist evolutionist who was converted to God and creationism.

Following is the conclusion to *Dry Bones ... and Other Fossils* :

And we hope that you, too, will try to see God's world through God's eyes. The heavens declare His glory; the fossils show the power of His judgment. And the open arms of Jesus hold us with the love of God that leads to abundant life forever for those who believe Him.
Thanks for sharing our adventures and thoughts on fossils. From our family to you and yours – May God bless you! (Parker, 1979, p. 71)

This is part of a *science course*. This is being read in the hundreds of thousands of copies sold in Australia and the United States. *This* is creation SCIENCE.
The excerpts above can only indicate in the slightest degree the overall classroom milieu. There are the eight filmstrips, the set of transparencies, guest speakers and, most important of all, a

committed creationist teacher popular with students. The only materials about evolution are those prepared by creationists, proving that evolution cannot be true.

In the seventh week of the course a test is given to prepare students for the final examination in the eighth week. By the seventh week, all the concepts have been well drilled. Students now realize that evolution and creation are *opinions*. There is no more evidence for one than for the other.

The final test is a condensed version of the pretest. Unless you knew what had occurred in the preceding weeks you could be excused for thinking that both evolution *and* creation had been taught. But the only thing taught about evolution is where it conflicts with creation. No evidence, just a handful of facts which give the appearance that both versions have been given equal time and equal effort.

The Final Vote

During the course the gifted children, who questioned the either-or or two-model presentation, were sent back to the library to view again the filmstrips. The authority of the printed word and the visual media are effective on all of us but much more so on 12-year-olds. The purpose of the whole course is to gradually lead students to the making of a decision and commitment which will affect their lives at the deepest level. The first step is to bring about acceptance that creation and evolution are equally valid models. The second step is to equate evolution with atheism. The final step is to evaluate the effectiveness of the indoctrination.

After the test the students were asked to vote creationism or evolution. Given the context of the vote and everything which had gone before, it is nigh miraculous that the six gifted children voted evolution knowing that by so doing they were voting atheist. Two actually became atheist and gave away religion.

It is appropriate to quote from Brother W. X. Simmons, director of the Catholic Education Office, Sydney. In November 1986 Simmons sent a letter to Catholic schools in Sydney which effectively banned the teaching of creation science. The letter also served as the foreword to *The Two Books of God*, a paper which gave the religious reasons for rejecting creation science.

Scientists do not accept creationism as science, and the Catholic student who is taught creationism as if it were a science is forced into an impossible position. No professional teacher should consider doing such. (Price, 1986b, p. A13.0)

The dilemma in which the students at Livermore were placed is a microcosm of the same dilemma in which many Christians have found themselves over the past century or more. The failure of the Christian churches to accept evolution because it appeared to contradict certain doctrines has caused many to conclude, as did the students at Livermore, that if the Church's science is so fallacious then so is its religion. This is one of the reasons why it may be said that Christianity is the religion which spawns atheism. The failure of Christian leadership to *proclaim* that evolution is acceptable means that for many the dilemma still exists. Thus Church leadership must bear at least part of the responsibility for the milieu in which creationism flourishes.

In the midst of the events at Livermore Chuck Larson, whose wife, Jan, was president of the Smith School Parent-Teachers' Association, appeared on public television. His words are a poignant comment on what has been said above. He says it all:

If evolution were to be absolutely, 100 percent proven to be true and that creationism was a lie, then, you know, what would I do with my belief as a creationist, as a Christian, if we go back to religion now? I would absolutely discard all of the Bible and everything in it. I, as a person, am a faulty person, and I have ignorances and faults and, you know, failures in my life, and I need an anchor and something to hang on to. And so if my God becomes a liar to me then He's no longer God, and I've got to look elsewhere for God. (Transcript of *Creation vs. Evolution, Battle in the Classroom,* KPBS Television, San Diego, 7 July 1982)

The Aftermath

The three or four mothers who approached the Livermore teacher to complain about their children being taught religion under the guise of science were ordinary suburban homemakers. In no way did they foresee that their action would catalyze the biggest scandal the nation had seen concerning the teaching of creation science in a classroom. They suddenly found themselves in the center of a storm. Protagonists from both sides rushed to Livermore to defend their interests.

Heavies from the ICR in San Diego arrived in town. Richard Bliss, an education expert, had with him Wendell Bird, the creationist lawyer who was to argue creation science in many a court, and still is. Attorney Bird advised Livermore creationists not to mention God when talking about Genesis (*Sunday Times*, California, 8 February 1981).

The upshot was that Livermore became a community split, a war zone with the bullets being fired by outsiders.

"Whether intentional or not, a child should never be forced to make a decision between God and science," said Mrs. Karlson, a former Catholic now attending a Presbyterian Church. "How many of us ever do that?"

After months of acrimonious debate and reams of publicity, the Livermore school board voted last month to suspend the teaching of scientific creationism by Ray Baird at the Emma Smith Elementary School.

The controversy, however, is far from over. There remains a passionate and powerful division of feelings – even though both camps claim to be unemotionally involved.

Neighbors are not on speaking terms even though they live on the same street, consider themselves Christians and have children who play on the same soccer team. (*The Herald*, California, 8 March 1981, p. 16)

A compromise was inevitable. The battle at the level of the school board seesawed back and forth. It wasn't much calmer in the Smith School Parent-Teachers' Association: the president was a creationist. For a while it looked as though Baird would be allowed to continue his course on the condition he consulted with parents and other members of the community.

One newspaper editor saw clearly enough through the cloud of emotion to get to the real issue:

CREATION SCIENCE IN THE CLASSROOM – A CASE STUDY

Ray Baird's Error

Ray Baird has admitted that he erred in bringing creationist materials into his classroom at Emma C. Smith School. He has promised not to use such materials again and to solicit community involvement for the creationist course he plans to teach in the spring.

Although Baird's admission of error was given with sincerity, and although he enjoys a widespread reputation as a competent and popular teacher, we believe that the error can't be dismissed so easily. The error was not a minor one, but involved an apparent violation of the First Amendment separation of church and state.

It also involved, in Baird's neglect to screen the creationist materials, a violation of one of the most fundamental of all the responsibilities expected of a professional teacher – one which demands special care when the subject is as sensitive as this one. Baird's belief in creationism is so fervent, it seems, that it rendered him incapable of maintaining his professional discipline.

We're profoundly concerned about his intention to teach the course again. We're convinced that regardless of how he dresses up creationism, the same problems will remain.

The issue, however, isn't really Baird. The issue is that teachers like him should be prevented from teaching this kind of subject in the future. Preventive guidelines are in order – specific ones concerning creationism and all other church-sponsored doctrines that purport to have scientific or academic legitimacy. If such guidelines aren't invoked, there are grounds to personally discipline Baird, and this should be done, as an example to discourage other teachers who might try to bring the church into the schools. (*The Independent,* California, 14 January 1981)

However, in the end, Baird's course was canceled and Baird went on to teach in a nearby senior high school.

A Conspicuous Absence

What is absent from the media accounts, all 12 square meters of cuttings when spread across the floor, is concern for the effect on the thirty students of Baird, and the students from the previous two years, and the students since. We can't help but wonder about them – what do they believe now, those who were forced to make such a monumental decision at the age of 12?

Did those students who chose the evolution horn of the false dilemma realize the truth later in their lives and become at ease with religion? Did some of those who chose the creationism horn realize later that it was scientific nonsense, and become atheists? Did those who chose creationism and stayed with it become, like Chuck Larson, fearful that if evolution were true then God must be abandoned?

Furthermore, how many Bairds in how many classrooms are re-creating the same dilemma? How many tens of thousands of children are being affected?

In Australia the Creation Science Foundation states on the envelopes it uses to mail its materials that the CSF is the "Supplier of quality Creation Science Material, Christian School Texts and Audio Visuals." The school materials listed in the catalogues are virtually the same as used at Livermore. The CSF sells at least a quarter of a million dollars' worth of texts and films each year. How many Australian parents who send their children to Christian schools are aware of what may be happening to those children – belief in God at the expense of the rejection of the scientific culture in which they live? Or the acceptance of science at the expense of faith?

There could be no more appropriate ending for this chapter than the words of one of the mothers whose children attended Baird's class. She still looks back with anger at what was inflicted upon her child.

I'll tell you honestly, that it infuriates me to think that some teacher, entrusted with the minds of gifted students, is feeding them twisted facts, gross exaggerations, and downright lies about the nature of the world because he feels it's his "duty" to save their souls (or whatever his particular problem is).

Chapter 13

Courts, Churches and Schisms

In the United States the Institute for Creation Research defends its nest at home and far away. It attacks the "enemy" whether it be a single teacher, a school board or the whole scientific establishment. The deepest condemnation, however, is for those it perceives as having betrayed the cause.

Early in 1987 every clergyman of the Christian Reformed Church received a malicious letter denouncing Doctors Young, Van Tell and Menninga, staff members of Calvin College in Grand Rapids, Michigan. Calvin College is supported by the Christian Reformed Church. The letter, in the form of a circular, was sent in the name of Duane Gish.

Dr. Young had been publishing and teaching that the earth was old. This was a courageous step in fundamentalist circles and has led to a shift to old-earth fundamentalism in those circles. Because there had been some press coverage the college conducted an investigation, apparently a mere formality. The three scientists seem to have been triumphantly acquitted.

In Australia the same has occurred. After his March 1988 debate with Gish in Sydney, Professor Ian Plimer, head of the geology department at the University of Newcastle, states that he received threats of legal action from creationist lawyer Wendell Bird on behalf of Gish and the Institute for Creation Research. In addition,

Professors Wieland and Rendle-Short, both of the Australian Creation Science Foundation, wrote to the Vice Chancellor of the University of Newcastle, demanding a public apology for Plimer's attacks on Gish during the debate. It seems that creationists will go to considerable lengths to harm the careers of those who disagree with them. Since then Professor Plimer has been informed by the Vice Chancellor that the University Council is unanimous in offering him their full support in his confrontation with creation science.

Textbooks With Blank Pages

In California the State Board of Education's antidogmatism policy states:

> That, on the subject of discussing origins of life and earth in public schools:
> (1) Dogmatism be changed to conditional statements where speculation is offered as explanation for origins.
> (2) Science should emphasize "how" and not "ultimate cause" for origins.

Mr Kelly Segraves has achieved notoriety by exploiting this law to the full. He chose Sandra Williams, his son's teacher of advanced biology at Serra High School in San Diego, as the victim he would offer as sacrifice. The class used a standard textbook, *Biology* by Helena Curtis. He charged her with breaking the state law and violating his son's constitutional rights by showing the students a segment from the world-renowned television series *Life on Earth* by Richard Attenborough. Segraves was director of the Creation Science Research Center and had previously filed a suit on behalf of his three children, challenging that teaching evolution contravened the State Board of Education's antidogmatism policy.

With the arrogance and confidence of the born fanatic, together with input on the side from creationist lawyers, Segraves dragged the unfortunate Williams through school committees, state committees, curriculum committees and just about every other committee for twelve months or more. That she was able to continue teaching through all of this is a tribute to her stamina and her courage.

Segraves listed all the laws Williams was breaking, and pointed out nearly two hundred "errors" in the textbook used by his son Kasey. These make interesting reading. Here are a few examples of these "errors":

- Evolution is made all important to biology (pp. 11 - 13).
- Science is equated with evolution. Modern creationist thought is ignored (passim).
- "Millions of years" is alleged (p. 370).
- The evolution of plants is asserted (p. 381).

The essence of his critique is that in a biology book evolution must not be asserted, nor must the word "evolution" ever be used. There must be no mention of anything older than 6000 years. Is there anything left in biology after the Segraves scissors have done their work?

Eventually Segraves sued the State Board of Education. In a judgment on 12 June 1981 Sacramento Supreme Court Judge Irving Perluss upheld the validity of the Board's 1978 policy document *Science Framework*. A 1984 addendum to the document mentions that

In directing the Board to disseminate its 1972 antidogmatism policy, the court was concerned primarily with the mode of presentation and not the content of the science curriculum. Thus it reaffirmed that science properly taught as the product of the scientific method is by definition not dogmatic in that it is open-ended and without preset conclusions. The court's concern was therefore directed toward ensuring that science not (be) presented as possessing absolute answers to questions. The court was particularly impressed with the testimony of a witness who emphasized the need for science educators to convey to the student's mind the clear understanding that science cannot exceed its own discipline and methodology.

But this is just one example of the tactics devised by Segraves and his mother, Nell. Since 1981 they have made into a fine art the harassment of those responsible for education, whether they be teachers, supervisors or entire boards.

In 1981 Nell organized the Creation Creed Committee, which

issued information kits to enable its hundreds of members – monitors? – to quickly sniff out any mention of the words "evolution" and "millions" in the science classrooms of California's 1100 school districts. One supposes that "millions" can still be used in mathematics classes. It is near the stage where the demand will be that all biology books with words in them are to be publicly burned!

The Creation Creed Committee is allied with Christian Voice, an organization claiming to represent 40,000 pastors. It also claims that of the thirty-six politicians placed on its hit list, twenty-three were defeated in subsequent elections (San Diego *Tribune*, 23 November 1981).

One wonders at the content of a biology course jointly constructed by Kelly Segraves and that teacher in Livermore, Ray Baird. Presumably Genesis is all that is needed.

But of course this can't happen in Australia. Or can it? In January 1988 thirty-five Christian schools in New South Wales bitterly opposed implementation of new state government legislation which forces them to teach evolution or be closed down.

> Mr Brian Wenham, of the Coalition Affirming Freedom in Education, said yesterday most schools would continue to teach creation – that the world was created in seven days – in divinity or religious classes instead of science (classes).
> Mr Wenham said most independent schools would obey the law and teach evolution in science classes. (*Courier Mail*, Brisbane, 7 January 1988, p. 4)

One wonders if parents with children at the other sixty or so Christian schools in Australia realize their children may not be taught evolution. Some may wish it that way. Others may not.

California Today, Tomorrow Texas

What the Segraves have done in California has been repeated all over the United States. Kelly Segraves and the Reverend Robert Grant have produced *A Manual for Pastors: Evolution and the Christian Child in California Public Schools*. This is published and distributed by Creation Creed Committees in conjunction with the

Christian Voice and National Heritage associations. This amazing thirty-page document is the distilled wisdom of the ten years Kelly Segraves spent in bringing school boards to total frustration and harassing teachers and principals to the point of despair. Every little loophole of the law and how to exploit it is explained. Even the wording to be used for messages over local radio is set out:

> Hello. I'm Dr. Robert Grant of the Creation Creed Committee. Let's face it, it's difficult enough for Christian children today, isn't it? They should not have to also tolerate the ridicule of their religious beliefs by teachers in the public schools. The California State Board of Education agrees with this. As you know, they recently clarified their policy. They said that evolution must not be taught dogmatically in the public schools.
>
> The Creation Creed Committee is a voluntary association of thousands of ministers and concerned Christian parents who are helping to assure that this policy is carefully observed in your child's school....
>
> This has been a public service announcement. Thank you very much.

In another age Segraves might have achieved the status of Inquisitor.

This could never happen in Australia – at least to that extent surely? But wait a minute. MACOS (Man, A Course of Study) was banned in Australia by conservative groups from Christian churches, including Catholic, about the same time it was being banned in some American states. British and Australian books on human biology were banned around the same time. One state has seen a married couple elected to Parliament after pledging to block legislation on moral issues they don't agree with. A nice lady on television was telling everyone why some books should be banned from schools because they didn't show the nice side of life.

Whenever an Australian television program on science or evolution mentions millions of years the station concerned is sure to receive an onslaught of protest letters. This must be having some effect. Recently one program described a project which will keep Australia at the forefront of investigations into deep space. A series of radio telescopes linked from northern New South Wales to Tasmania, a distance of about 2000 kilometers (1240 miles), will enable the probing of deep space through billions of light-years. Apparently forbidden to mention any age ending in -ion, the announcer said the project would investigate space *dozens* of light-years away, a triumph for the CSIRO designers of the system. The announcer played it safe and kept well and truly under the dreaded 6000-year barrier.

For all that, never will Australia see the situation which has occurred in Texas. In the one town at the same time each year fifty million dollars are spent over the space of a week buying textbooks for every single child in the state! Imagine what would happen if the creationists got control of that system.

Ah, but they did. For the decade up till 1985 evolution was virtually eliminated from schoolbooks in Texas. Publishers couldn't afford to do otherwise.

> For example, Coronado, the publisher of *World History: A Basic Approach* (1984), was asked to change "The earliest people lived on Earth over 1 million years ago" to "Many scientists believe that the earliest people lived on Earth over 1 million years ago."
>
> In another case publishers of two dictionaries being considered for high school use were asked to delete "objectionable" words. Houghton Mifflin, publisher of *The American Heritage Dictionary*, agreed to made the changes, while G. & C. Merriam, publisher of *Webster's New Collegiate Dictionary*, in a strongly worded letter to the commissioner of education, refused to censor the "objectionable" words. As a result, Texas adopted no dictionaries in 1981, since it requires at least two listings for each subject. (Moyer, 1985, p. 23)

In addition, because publishers maintained they couldn't afford to publish one version for Texas and another for the rest of the country, what Texas wanted the rest of the United States got too. These tactics, masterminded by Mel and Norma Gabler, have adversely affected science teaching throughout the country.

You must agree: *that* could not happen in Australia. But what about this?

Ray E. Martin is an American creationist who obviously scorns the theoretical approach. He attacks evolution in the most practical way. No dreary lawsuits or time-consuming harassment of school boards for Ray. He has adopted his own original approach and he is not afraid to tell the world. Here is an excerpt from an article he wrote, aptly titled "Reviewing and Correcting Encyclopedias":

Encyclopedias are a vital part of many school libraries ... (They) represent the philosophies of present-day humanists. This is obvious by the bold display of pictures that are used to illustrate paintings, art, and sculpture ... This makes it important that the materials we place before our children are free from ... that which would enflame passions.

(In encyclopedias) we are not battling a plot that captivates minds but are looking for erroneous information, sensual pictures, and unchaste details ... One of the areas that needs correction is immodesty due to nakedness and posture. This can be corrected by drawing clothes on the figures or blotting out entire pictures with a magic marker. This needs to be done with care or the magic marker can be erased from the glossy paper used in printing encyclopedias. You can overcome this by taking a razor blade and lightly scraping the surface until it loses its glaze. After this is done the magic marker will not erase.

(As for evolution,) ... cutting out the sections (on the subject) is practical if the portions removed are not thick enough to cause damage to the spine of the book as it is opened and closed in normal use. When the sections needing correction are too thick, paste the pages together being careful not to smear portions of the book not intended for correction.

Reproduction: Two basic categories of reproduction are human and animal. Reproduction in the animal kingdom shifts the attention away from the human level enough to make it a more neutral platform for the presentation of facts

... (As for human reproduction,) volume description and page numbers can easily be passed from one student to another, and an atmosphere may begin taking shape ... that will lead to a breakdown in moral reserve and purity of thought ...
The work (of cutting out or pasting together pages) will be done by those who have a burden for that which is right ... school board members and the pastoral committee ... burdened parents ...
The work is great and the need is urgent. Who will rise to the task? Will we respond as Isaiah: "Here am I; send me"? (Quoted by Marty, 1983, p. 86)

Despite the similarity in names Martin and Marty are very much on opposite sides.

After receiving a copy of this quote, the librarian at the University of Newcastle checked through some books, to find that every reference to evolution had been cut out from books in the paleontology section of the university library. Other Australian university libraries are investigating whether the same has occurred in their libraries.

See You in Court

In Australia a perennial topic of conversation is how a handful of bloody-minded unionists, particularly in demarcation disputes, can bring the country to a halt. We have come to accept it as a way of life. People in other countries think we're crazy. It's part and parcel of being a democracy, we say. It's just the British disease, we say.

But the havoc a handful of organizations and a dozen or two individuals have been able to wreak through tapping into legislative processes in the different American states is mind boggling.

The Australian situation is different because Australia doesn't have a first amendment to the Constitution which guarantees the separation of church and state. Now it is not the intention to give a blow-by-blow description of the many court cases which have ensued in America because of that amendment. That is properly the task of a historical study and not of relevance to this book, fascinating though it may be. What is relevant is a selective look at some of the briefs in the two major court cases to date: the Arkansas decision in 1981 and the U.S. Supreme Court decision in 1987.

Until 1970 creationists had been attempting, with little success, to have evolution removed from the school curriculum. In 1970 a new tactic was employed which met with immediate success and changed the whole ball game. Nell Segraves heard about a court case in which an atheist mother successfully sued in court to have her child exempted from prayers at school. Inspired by this she fought school boards and courts to have her grandchildren exempted from anything resembling evolution. The success of the Segraves in this has been described earlier.

In 1979 Paul Ellwanger drafted a model bill which, if passed by state legislative bodies, would ensure that creation science and evolution would have equal time in science classes. At first creationists went around collecting signatures for a referendum but found to their surprise that they could not get enough numbers. Ellwanger then adopted the tactic of persuading sympathetic senators to introduce the bill in the course of normal legislation. This was done successfully in Arkansas in 1981, appealed against in the same year and ruled against early the following year.

In terms of evidence and documentation the Arkansas case produced two important documents which highlight the nature and tactics of creation science. The first is Act 590 and the second is the decision by Judge Overton. They are extremely useful for any organization or institution involved in legislating against creation science, or even just for understanding what creation science is all about.

Act 590

Act 590 of 1981, State of Arkansas, 73rd General Assembly, Regular Session 1981 became known simply as "Act 590" or "Balanced Treatment for Creation-Science and Evolution-Science Act". It was signed as law on 19 March 1981 by Frank White, governor of Arkansas. The act contains virtually every argument ever used to prove that creation science and evolution are on an equal footing as science.

The definitions of evolution science and creation science do not use the word "creator" or "god." They are contrasted with each other on six points: origin of universe, evolution of life, biological evolution of species, ancestry of man, geology of earth, and age of earth. There are twelve "Legislative Findings of Fact"; here are three examples:

> 7(c) Evolution-science is not an unquestionable fact of science, because evolution cannot be experimentally observed, fully verified, or logically falsified, and because evolution-science is not accepted by some scientists.
> 7(g) Public school instruction in only evolution-science also violates the principle of academic freedom, because it denies students a choice between scientific models and instead indoctrinates them in evolution-science alone.
> 7(j) Creation-science is an alternative scientific model of origins and can be presented from a strictly scientific standpoint without any religious doctrine just as evolution-science can, because there are scientists who conclude that scientific data best support creation-science and because scientific evidences and inferences have been presented for creation-science.

Section 7(c) states that evolution is not an unquestionable fact of science because it cannot be logically falsified. Arguments using this definition of science to support creationism have appeared in both Australian and American academic journals. These writings are based on the writings of philosopher Sir Charles Popper, who stated that neither evolution nor creationism can be validated so they are both hypotheses. At the time he wrote this Popper was avowedly anti-evolution. His achievement in other areas carried weight. What none of the papers using Popper to justify creation science in schools remembers is that as far back as 1972 Popper recanted his position:

> I blush when I have to make this confession; for when I was younger, I used to say very contemptuous things about evolutionary philosophies. When twenty-two years ago

Canon Charles E. Raven, in his *Science, Religion, and the Future*, described the Darwinian controversy as a "storm in a Victorian teacup," I agreed, but criticized him for paying too much attention "to the vapors still emerging from the cup," by which I meant the hot air of the evolutionary philosophies (especially those which told us that there were inexorable laws of evolution). But now I confess that this cup of tea has become, after all, *my* cup of tea; and with it I have to eat humble pie. (Cole, 1981, p. 44)

The Churches Have a Say

The appeal against Act 590 was heard by U.S. District Court Judge William R. Overton between 7 December and 17 December 1981. The judge then took two weeks to write up his decision. It is clearly written and comprehensive, and it covers evidence from a number of areas. One gets the impression that there was no way in the world that the slightest chink would be left open to enable an appeal. On 5 January Judge Overton handed down his decision. He upheld the appeal and ordered costs against the State of Arkansas.

What must have staggered the creationists was that almost all the plaintiffs were clergy. They included the resident Arkansas bishops of the United Methodist, Episcopal, Roman Catholic and African Methodist Episcopal churches, the principal official of the Presbyterian churches in Arkansas, and other United Methodist, Southern Baptist and Presbyterian clergy. As well, the American Jewish Congress, the Union of American Hebrew Congregations and the American Jewish Committee sued on behalf of members living in Arkansas. The only scientific plaintiffs were a science teacher and the National Association of Biology Teachers.

Langdon Gilkey was a witness at the trial. In 1985 he published a book describing his experiences, *Creationism on Trial*. He tells how the Methodist bishop of Arkansas, the Reverend Kenneth Hicks,

stated how important it was to the churches of Arkansas, and to his own large denomination (eight hundred churches in Arkansas), that the separation of the authority of the

State from the life and affairs of the churches be maintained, especially in the public schools. The Methodist denomination, and many of those other denominations for which he was privileged to speak, had been established on the principle of freedom of religion from State control, and they would struggle long to maintain that freedom. (Gilkey, 1985, p. 83)

Following the trial, some of the churches published affirmations of the result of the trial. The *1983 Minutes of the 195 General Assembly of the United Presbyterian Church*, which coincided with the 123rd General Assembly of the Presbyterian Church in the United States, urged pastors and congregations to study and discuss the decisions of the trial and the issues in the creation science controversy. They urged "pastors and Christian educators to help their congregations to interpret the biblical passages dealing with creation and the origin of human life in ways that take their message seriously" (p. 682).

In May 1983 the United Church Board for Homeland Ministries stated in a position paper, *Creationism, the Church, and the Public School* (Sec. II):

1) We testify to our belief that the historic Christian doctrine of the Creator God does not depend upon any particular account of the origins of life for its truth and validity. The effort of the creationists to change the book of Genesis into a scientific treatise dangerously obscures what we believe to be the theological purpose of Genesis, *viz*., to witness to the creation, meaning, and significance of the universe and of human existence under the governance of God. The assumption that the Bible contains scientific data about origins misreads a literature which emerged in a pre-scientific age.

2) We acknowledge modern evolutionary theory as the best present-day scientific explanation of the existence of life on earth; such a conviction is in no way at odds with our belief in a Creator God, or in the revelation and presence of that God in Jesus Christ and the Holy Spirit.

There is no better statement on the point at issue than this.

Tactics Used to Pass the Act

Part II of Judge Overton's decision describes how Senator James L. Holsted introduced the act:

> Holsted, a self-described "born again" Christian Fundamentalist, introduced the act in the Arkansas Senate. He did not consult the State Department of Education, scientists, science educators or the Arkansas Attorney General. The act was not referred to any Senate committee for hearing and was passed after only a few minutes' discussion on the Senate floor. In the House of Representatives, the bill was referred to the Education Committee which conducted a perfunctory fifteen minute hearing. No scientist testified at the hearing, nor was any representative from the State Department of Education called to testify. (Overton, 1981, p. 937)

The person who persuaded Holsted to introduce the act was the Reverend Mr. Blount, at the instigation of Paul Ellwanger. Overton points out that Act 590 is identical with Ellwanger's model bill except for minor typological changes.

The Origin of Act 590

Overton clearly states how the act was jointly framed by Ellwanger and Wendell Bird, the same lawyer who came post-haste to Livermore when that fracas erupted:

> Citizens For Fairness In Education is an organization based in Anderson, South Carolina, formed by Paul Ellwanger, a respiratory therapist who is trained in neither law nor science. Mr. Ellwanger is of the opinion that evolution is the forerunner of many social ills, including Nazism, racism and abortion ... About 1977, Ellwanger collected several proposed legislative acts with the idea of preparing a model state act requiring the teaching of creationism as science in opposition to evolution. One of the proposals he collected was prepared by Wendell Bird, who is now a staff attorney

for (the Institute for Creation Research). From these various proposals, Ellwanger prepared a "model act" which calls for "balanced treatment" of "scientific creationism" and "evolution" in public schools. He circulated the proposed act to various people and organizations around the country. (Overton, 1981, p. 936)

Overton then quotes from letters which Ellwanger had sent to various people and which showed quite clearly that neither Bird nor Ellwanger himself believed in the slightest that creation science was science. Their private beliefs were quite contradictory to the sum total of all the arguments they had devised for their Act 590. Overton also quotes from other letters urging ministers of religion "not to be apparent or in the forefront of those who publicly advocate the implementation of Act 590.

Overton finds "most interesting" Ellwanger's own testimony concerning his strategy:

(Overton) You're trying to play on other people's religious motives.
(Ellwanger) I'm trying to play on their emotions, love, hate, their likes, dislikes, because I don't know any other way to involve, to get humans to become involved in human endeavors. I see emotions as being a healthy and legitimate means of getting people's feelings into action, and ... I believe that the predominance of population in America that represents the greatest potential for taking some kind of action in this area is a Christian community. I see the Jewish community as far less potential in taking action ... but I've seen a lot of interest among Christians and I feel, why not exploit that to get the bill going if that's what it takes.

Overton continues at some length to discuss distinctions between science and creation science before concluding that creation science is not a science but a religion.

One creationist witness was Dr. Wickramasinghe, a scientist who believed that the genetic material which "seeded" life on earth came from outer space. This is not an uncommon theory nowadays but one can only assume that creationists thought the scientist was agreeing with Henry Morris's well-known statements that the marks on Mars and the moon are leftovers from the primeval battle between the devil and the good angels, and that UFOs are "manned" by the evil spirits who dwell in the distant planets and the asteroid belt. Apparently the creationists misunderstood the witness. There is more than a hint of humor in Overton's statement:

> The Court is at a loss to understand why Dr. Wickramasinghe was called in behalf of the defendants. Perhaps it was because he was generally critical of the theory of evolution and the scientific community, a tactic consistent with the strategy of the defense. Unfortunately for the defense, he demonstrated that the simplistic approach of the two model analysis of the origins of life is false. Furthermore, he corroborated the plaintiffs' witnesses by concluding that "no rational scientist" would believe the earth's geology could be explained by reference to a worldwide flood or that the earth was less than one million years old. (Overton, 1981, p. 940)

According to Gilkey's book on the trial there were more than a few moments of humor. At one stage, when a creationist was explaining how creation science was censored in schools, it was pointed out as an example that social studies texts which discussed the transition of man from nomadic hunters to farmers left out Cain's theory! Cain was the son of Adam and Eve who killed his brother Abel. The Bible does not mention him as having any particular theory.

Impossible to Teach

One of the important aspects of the trial for schools is the evidence which showed conclusively that it is impossible to construct a science course from creationist materials:

> The testimony of Marianne Wilson was persuasive evidence that creation science is not science. Ms. Wilson is in charge of the science curriculum for Pulaski County Special School

District, the largest school district in the State of Arkansas. Prior to the passage of Act 590, Larry Fisher, a science teacher in the District, using materials from the (Institute for Creation Research), convinced the School Board that it should voluntarily adopt creation science as part of its science curriculum. The District Superintendent assigned Ms. Wilson the job of producing a creation science curriculum guide. Ms. Wilson's testimony about the project was particularly convincing because she obviously approached the assignment with an open mind and no preconceived notions about the subject. She had not heard of creation science until about a year ago and did not know its meaning before she began her research.

Ms. Wilson worked with a committee of science teachers appointed from the District. They reviewed practically all of the creationist literature. Ms. Wilson and the committee members reached the unanimous conclusion that creationism is not science; it is religion. They so reported to the Board. The Board ignored the recommendation and insisted that a curriculum guide be prepared.

In researching the subject, Ms. Wilson sought the assistance of Mr. Fisher who initiated the Board action and asked professors in the science departments of the University of Arkansas at Little Rock and the University of Central Arkansas for reference material and assistance, and attended a workshop conducted at Central Baptist College by Dr. Richard Bliss of the ICR staff. Act 590 became law during the course of her work so she used Section 4(a) as a format for her curriculum guide.

Ms. Wilson found all available creationists' materials unacceptable because they were permeated with religious references and reliance upon religious beliefs. (Overton, 1981, p. 940)

Marianne Wilson was forthright in describing the curriculum materials given to her by Richard Bliss, curriculum director of the Institute for Creation Research. This was reported in *Science*, the most prestigious science journal in America:

"What he had was trash," says Wilson. "It was just full of religious references, and the science was awful."
Eventually Wilson and Fisher did put together a curriculum that has noncreationist material as references. However, very little of the material comes from conventional scientific sources, and one article referred to is in *Reader's Digest.* (*Science*, 29 January 1982, p. 486)

On the same page of *Science* it was reported how, in a roundabout way based on the American version of the old boys' network, Governor Frank White had been approached and agreed beforehand to sign the bill if it came to his desk.

The Arkansas decision was not appealed against by creationists in a higher court. One reason may have been the comprehensive way in which Judge Overton wrote his brief. Another reason may well have been that the governor would not be likely to receive much support from a bureaucracy who woke up one morning to find that an act had been implemented of which they knew absolutely zero. Without an appeal heard before the U.S. Supreme Court the decision would apply only in Arkansas. But there was not too long a wait.

Louisiana

A few months after Governor White signed Act 590 in Arkansas a bill similar to Act 590 but with some minor changes was enacted in Louisiana, on 21 July 1981. In 1985, U.S. District Court Judge Adrian Duplantier declared the law unconstitutional following a suit filed against the State of Louisiana in 1981. This was affirmed in 1983 by a three-judge panel, and in 1985 the Court of Appeals declined by an 8 to 7 vote to rehear the case. On 19 June 1987 the U.S. Supreme Court handed down a 7 to 2 verdict which agreed with the decision of the two lower courts that Louisiana's creationism act was unconstitutional because it had a religious purpose, not a secular one. This was a stinging defeat for creationists, especially attorney Wendell Bird who led the appeal team.

It is argued that the reversal inflicted at the federal level will simply drive creationists back to the grassroots level where they have best proved their effectiveness. One outcome of the Supreme Court decision was that it occasioned a blistering attack on Henry Morris and Duane Gish from another creationist group.

Who Is the Heretic?

Tom McIver (1988, pp. 10-11) describes the 1987 Creation Fellowship Conference in Baltimore, where John Robbins delivered a paper titled "The Hoax of Scientific Creationism." He says that "in a century of religious and scientific hoaxes, scientific creationism may be the biggest." He states that creation science is a "fraud and deception" and he is appalled that creation science lawyer Wendell Bird attempted to pass off creation science as science. It is a "shallow devious tactic doomed to failure."

Robbins went on to say that Henry Morris's contention that creation science need not contain any concept of God or the book of Genesis is "conning Christians into supporting a movement that is betraying its very principles." He asked Christians whether they should continue to spend thousands of dollars supporting a method of defending the faith "which inexorably leads to non-Christian conclusions."

Morris, in reply, did not deny the charges Robbins made, but his defense was that creation science was "the only version which would be allowed in public schools."

The briefs submitted to the Supreme Court were from as wide a range of clergy and organizations as in the Arkansas trial. The American Council for Civil Liberties also prepared a brief. By far the most impressive was the brief presented by seventy-two American Nobel Laureates in science. It may be worth mentioning that the references cited in the creationist brief did not include a single book by a creationist writer. In the brief from the Nobel Prize winners the majority of books cited were by leading creation science writers: two by Gish, seven by Morris, two by Parker, two by Bliss and one by Snelling (Klayman *et al.*, 1986, p. 27).

And all this brings us to the analysis of two distinct groups of creationists which are split by different interpretations of the last book in the Bible. This will complete the overall picture of creationism.

Pre-Millennium or Post-Millennium?

The millennium is a concept developed from the last book in the New Testament, the Revelation to John. This book epitomizes apocalyptic literature, a type of writing that is coded and highly symbolic. The literal reading of Revelation Chapter 20, verses 1 to 5, is that an angel will come down to earth and chain up Satan for a thousand years so that he cannot deceive the nations during that time. Those who have been executed for Christ and those who have not worshipped the Beast will come to life and rule as kings with Christ for a thousand years. This is the first raising of the dead.

Creation science and most creationists are pre-millennialists, or pre-mills. Many believe the Second Coming of Christ is just around the corner. Post-millennialists, as the name implies, believe the Second Coming has already occurred and the world is in the thousand-year time span. Post-mills accuse pre-mills of relying on scare tactics in order to convert people to Christ so they won't go to hell, of pessimism concerning political involvement and of "not being concerned with building a totally Christian society." They charge that the pre-mill attitude is "ineffective for promoting critical attitudes, including creationism" (McIver, 1988, p. 10). Many pre-mills believe we are living in the Last Days and they actually welcome the idea of Armageddon being so near.

Pre-mills accuse post-mills of emphasizing worldly success and materialism. The post-mills believe that the millennium will be established as a result of the victory of Christianity in this world. Post-mill is essentially optimistic while pre-mill is a pessimistic attitude to this worldly life.

To confuse the matter further, there are also a-millennialists, who claim the millennium is purely symbolic, as well as a number of other groups who have beliefs that seem to be a mixture of pre- and post- and a-millennialists.

It is not possible to give an extended treatment of the two main types in a short space but it is possible to state a few conclusions of the post-mills which are likely to result in attempts to change the fabric of democratic society. McIver (1988) should be consulted for a more comprehensive view.

Post-mills believe we should still be subject to Old Testament law, that biblical principles should apply to government and that Christians should get involved in government in order that Christianity will achieve dominion over mankind as well as creation. Post-mills believe that if Genesis contradicts science and history – as it does – then Christians should commence the task of reconstructing science and history to make them fit Genesis. Our presupposition that the Bible is true means that we work from there. This is extended to all walks of life – arts, science, politics ... the lot.

Because there is a return to Old Testament law, there must be a state with no distinction between religion and government. Democracy is a heresy and theocracy is the form of government ordained by God. All men are not created equal and the notion of human rights was introduced by Satan in the Garden of Eden.

These are a mixture of the beliefs of a cross-section of post-mills and do not equally apply to any one group. They are intended to give a picture, a feeling, that the other side of creation science may be even darker and more dangerous.

The "lying for God" admitted to by some creationists may be a tiny fib compared to what is proposed by post-mill reconstructionist Gary North:

> As a tactic for a short-run defense of the independent Christian school movement, the appeal to religious liberty is legitimate. Everyone who is attempting to impose a world-and-life view on a majority (or a ruling minority) always uses some version of the liberty doctrine to buy himself and his movement some time, some organizational freedom, and some power ... So let us be blunt about it: we must use the doctrine of religious liberty to gain independence for Christian schools until we train up a generation of people who know that there is no religious neutrality, no neutral law, no neutral education, and no neutral civil government. Then they will get busy in constructing a Bible-based social, political, and religious order which finally denies the religious liberty of the enemies of God. (McIver, 1988, p. 15)

Gary North is one of twenty-two speakers mentioned in a pamphlet advertising the July 1986 Continental Congress on the Christian Worldview III, held in Washington. The conference was sponsored by the Coalition on Revival (COR).

The background to the conference is given under the heading *When in the Course of Human Events ...*, the opening words of the Declaration of Independence. The tone of the conference may be judged from some excerpts from the pamphlet:

In His inerrant, written Word, God stated for all time how all men ought to live, govern themselves and conduct their affairs. The Coalition on Revival was called into existence ... to cause "God's will to be done on earth as it is in heaven."

The rest is quoted in full:

These 17 documents state the Biblical World View of Law, Government, Economics, Business, Education, Arts and Communication, Medicine, Psychology and Counseling, Science and Technology, Christian Unity, Evangelism and Missions, Discipleship, Helping the Hurting, Social and Political Moral Issues, Revitalizing Christian Colleges and Seminaries, The Family and Pastoral Renewal.
At the Continental Congress On The Worldview July 2-4, 1986 in Washington D.C., these Worldview Documents and the Manifesto For The Christian Church will play a modern role analagous to Luther's 95 Theses, the Westminster Confession and the Declaration of Independence and, as such, will be "nailed to the Church's Door" by 60 of COR's Steering Committee members and their regional editorial committees. A Nehemiah type Solemn Assembly and signing ceremony will take place July 4th at the Lincoln Memorial. There participants will make a covenant with God and with each other to live in obedience to the Bible until they die. By God's grace, this event and the unleashing of these documents onto the Church and the world will mark a turning point in history wherein the Body of Christ will

awaken and take its proper leadership role with the world. Then, once again, we can see the Church transforming the world and influencing it to conform to Biblical standards rather than continuing to allow the world to conform the Church to its standards.

The only thing left unsaid is who is going to be the boss after this takeover of the world. But it does not take a giant leap of the intellect to guess the answer.

The title of Gary North's address to the conference is "Why God has waited for world revival until now." Is this, as the title suggests, a direct message from above? North is listed as president, Institute for Christian Economics, so presumably he is a hot contender for the upcoming office of World Treasurer. One wonders what alterations will be needed to the U.S. Constitution and the Bill of Rights.

Each of the twenty-two speakers is identified by a small photograph. The titles of the speeches are just as revealing as the publicity blurb: Colonel Doner, "A Practical Strategy for Repossessing Our Government," R. J. Rushdooney, "Christian Conquest of the World," Robert Weiner, "World Dominion Through God's Glorious Church."

A key speaker is, of course, Duane Gish on "The Scientific and Biblical Evidence for Creation." Without Gish's evidence the whole edifice of God's inerrant word would crumble. Hence it is of absolute necessity that the fact of evolution be kept at bay.

However, the surprise of the conference, or maybe not, is the address "America's Return to the Spiritual Values of the Founding Fathers" by Ronald Reagan, President, United States of America.

Perhaps a trivial point to the conference but one which is of some interest to those of us who live in other lands is: What happens if we don't want to be dominated by a Christian World Government?

The Moral of the Story

There is an urgent lesson in this for Australia. We are partway along the path to the situation in which the United States now finds itself. The separation of church and state in the U.S. Constitution is obviously insufficient protection against groups determined to undermine public school education as well as any other form of education which does not toe their party line.

It is apparent that modern teaching on evolution, appropriate to the age of the student, must be a prerequisite in all school science courses at the junior level and not, as is mostly the case at present, only in senior level biology classes. If evolution is a central concept which permeates every branch of modern science then it must be seen as such in beginning science courses. This will require additional resource materials for students and teacher inservice training.

Evolution as fact is necessary to protect children against the poison of creationism but, by itself, it is not sufficient. In the area of humanities the study of religion as a human phenomenon is equally necessary. Without this there is a yawning gap in a child's education since that aspect of the human being we label as "religious" will be filled in one way or another, whether by sects, cults, superstitions, astrology or creationism, or whatever.

In the past, such courses in schools have been opposed by some major church traditions because they were seen as perhaps leading students to conclude that their particular religion did not have all the answers. Those churches committed to students being made aware of only one narrow doctrinal belief system still oppose such an approach. The skills acquired by students in analyzing religion as a phenomenon are equally applicable to quasi-religious belief systems such as communism or secular humanism.

The academic study of religion most often uses a so-called typological approach or some variation thereof. In South Australia the Education Department has completed and trial-taught such a course from Kindergarten to Year 12 for use in public schools. It is also being used in some church schools. Other such courses exist to a greater or lesser extent in other states.

To make just one illustration. The study of the creation stories across many cultures throughout the world, including tribal religions, indicates clearly to anyone that Genesis, in common with all other such stories, is a myth used as a vehicle for religious truth —not meant to be taken literally. As such, it is a more powerful weapon against creationism than science. Moreover, in common with other creation stories, it addresses itself to the perennial questions of the self and its relation to the universe, the same questions which have always been asked. Seeking the answers to such questions is part of being human, and this human search finds expression in seeking to know the nature of reality. This is what we call science.

It goes without saying that the acquisition of such knowledge in schools is anathema to creationism.

Chapter 14

Australian Creationism

Until 1970 American creationists had campaigned for the removal of evolution from school science curriculums. In this they had little success. The early 1970s witnessed the initiation of new tactics which in the 1980s resulted in a significant weakening of school science courses across the nation.

The strategies used, while logically contradictory, have been highly effective. The wide front of the creationist tactics includes, first and foremost, a determined effort to produce a credible public image, to show the public that creation science in its own right is as equally valid an explanation for origins as is evolution. This image bolsters *the creationist demand for equal time in school science courses.*

Running parallel has been the tactic of weakening, or removing entirely, references to evolution in textbooks on the grounds that it is offensive to the religious beliefs of some pupils.

Coupled with both of these tactics has been the demand that creation science has a right to be taught in science courses in the name of academic freedom. Creationists found it unacceptable that creation science be relegated to some other area of the school curriculum, such as general studies. Documents are extant which detail the arguments above and also how to implement them with politicians, education departments, schools, teachers and parents. Underpinning an organized and cohesive effort at every stratum of the educational enterprise is ceaseless activity at the grassroots level.

For example, a couple of months after the U.S. Supreme Court had ruled creation science a religion, Paul Ellwanger had constructed a package titled *A Uniform Origins Policy*, subtitled *A Policy for Academic Freedom in Origins Teaching (For State Legislatures and Boards of Education)*. The policy is described in the news release which is part of the package:

> The policy does not mention any specific origins concept. Nowhere in the policy are the words "evolution" or "creation" used. A policy does not give a legislative mandate. This policy recommends and encourages. It also has some built-in features for teacher employment protection. (KDF, 1988, p. 5)

The document contains not-so-veiled threats of legal action if authorities take action against those teaching creation science in the classroom:

> (b) If a teacher is terminated, suspended, or otherwise disciplined after attempting in good faith to carry out the intent and provisions of this policy, there shall be a rebuttable presumption that the termination, suspension, or discipline of the teacher was an unlawful retaliatory measure which injures the rights of academic freedom enjoyed by public school teachers in this jurisdiction. (KDF, 1988, p. 5)

There is little doubt that "academic freedom" will also become a catchcry in Australia for those demanding a place for creation science in the school science curriculums.

The Creation Science Foundation

The foundation celebrated its tenth anniversary in 1988. Its origins lie in the amalgamation of two pre-existing groups – one in Queensland led by Ken Ham and John Mackay, the other in South Australia led by Dr. Carl Wieland, a medical doctor who had begun publishing *Ex Nihilo* magazine in June 1978. The inspiration for the formation of the Creation Science Foundation in 1980 followed upon the visit to Australia of two lecturers from the Institute for Creation Research in San Diego, Dr. Harold Slusher and Dr. Richard Bliss.

The Creation Science Foundation (CSF) in Queensland is closely modeled on the Institute for Creation Research (ICR) in California. There has been considerable input from the "mother house," particularly in the early stages when nearly all materials marketed by the CSF were from American sources. Lecturers from the ICR are frequent visitors to Australia. Ken Ham, a founding member of the CSF, has spent time in the United States producing a set of films presumably to replace the *Origins* series of six films which are marketed by, but not owned by, the two creation science groups. This joint venture has a huge potential market. Ham has stayed on in the United States as a staff member of Christian Heritage College and is now joined with other ICR staff on the debating circuit.

The Aims of the Creation Science Foundation

The following quote is prominent in the *Ex Nihilo Technical Journal*:

What is the Creation Science Foundation?
The Creation Science Foundation Ltd., is an independent, non-profit, non-denominational organization, controlled by Christian men and women of science and education, committed to researching, developing, and promoting Christian creationist materials, and Christian school texts and aids. Our work is based on acceptance of:
1. The Bible as the divinely inspired written Word of God. It is the supreme authority in all matters of faith and conduct.
2. The final guide to the interpretation of Scripture is Scripture itself.
3. *The account of origins presented in Genesis is a simple but factual presentation of actual events and therefore provides a reliable framework for scientific research into the question of the origin and history of life.*
4. The scientific aspects of creation are important but are secondary in importance to the proclamation of the Gospel of Jesus Christ.

5. The Doctrines of Creator and Creation cannot ultimately be divorced from the Gospel of Jesus Christ.

Please note that in all of this, we do not hide the fact that our work is centered around Jesus Christ. We are convinced that the real needs of men and women can only be met when individuals are restored to a personal friendship with Jesus Christ the Creator. Such a friendship can only be achieved by personal acceptance and commitment to Jesus Christ who is God the Creator. (Snelling, 1984, inside front cover)

Thus, for creationists, an important, if not *the* most important, aspect of creation science is its use as a *tool for evangelism.*

In their court cases and public utterances creationists consistently claim that their science is not dependent on the first eleven chapters of Genesis. They imply that the 2 million or so scientists from every nation who have accepted evolution, most of whom have probably not read the Bible, would have arrived at six-day creation, Flood geology, and so on, if they hadn't been diverted by Satan. Thus creationists see no need to hide their religious motivation, although they prefer to preserve the public image that they are an alternative honest group of scientists whose theory of origins just happens to coincide with the biblical account.

This is quite clear in point 3 of the CSF's statement above. Henry Morris's statement, made after creation science was labeled religion by the U.S. Supreme Court, is an obvious indication that the leaders of the movement are clear in their own minds that they are living with a lie. Morris said, in effect, it was necessary to *pretend* that creation science is science and not religion *because that was the only way there was any chance of getting it into schools.*

Membership of the CSF

The Creation Science Foundation was formed in 1980 with just seven members and seven directors. The seven members were also the seven directors. At annual general meetings thereafter the same seven members continued to elect themselves as directors. No financial reports have ever been issued to the many thousands of subscribers who support the CSF.

The CSF is a tax-exempt company, and by law such a company

must furnish audited statements of its financial activities, membership and directors to the Corporate Affairs Commission on 31 March of each year. In 1980 these records show the seven director-members as:

David John DENNER	Teacher
Robert Stephen GUSTAFSON	Solicitor
Kenneth Alfred HAM	Missionary
John Barry MACKAY	Missionary
Alfred John Maynard OSGOOD	Medical Practitioner
Tyndale John RENDLE-SHORT	Medical Practitioner

Secretary & Director:
John Andrew THALLON   Accountant

The auditors listed are Peat Marwick Mitchell, and John Thallon doubles as director and secretary. Ham and Mackay were at the time high school teachers but listed themselves as missionaries.

Thallon continued as director, secretary and member until 1985, since when there has been no mention of him in any of these positions. Company records dated 10 October 1984 disclose:

The company has entered into a contract with Tralil Pty. Ltd. for the provision of management consulting services for the period 1st September, 1984 to 30th June, 1985. John Andrew Thallon declared at the Executive Committee Meeting, 27th September, 1984, that he is a director of Tralil Pty. Ltd. which is the trustee for a trust of which he is a beneficiary.

This contract with Tralil is presumably a result of investment losses noted in the *Statement of Income and Expenditure for the year ended 31st March 1985*, which records "Extraordinary Item Loss of Investments, 1984, $47,939 and 1985, $44,424."

David Birt Bardsley, who gives his occupation as "administrator," appears as secretary on a list of directors dated 30 July 1986. The same list records the joint resignation on 30 July 1986 of two of the founding directors, David John Denner (teacher) and Robert Stephen Gustafson (solicitor), although in 1988

Gustafson still acted as the foundation's director. The directors are now down to four out of seven, since Thallon is no longer listed as a director. Ham, Mackay, Osgood and Rendle-Short are those who remain. All are founding directors.

However, there are more changes to come. Less than two years after his appointment as secretary, David Birt Bardsley is "removed" on 7 October 1987 and Gregory Ross Peacock takes over as the new secretary on 13 October 1987. Peacock had been admitted as a director six months previously.

Prior to all this, on 4 April 1985 the auditors (Peat Marwick Mitchell) notified the Corporate Affairs Commission of their resignation on 25 March 1985, a most unusual step since the company returns were due one week later on 31 March. On 18 September 1985 a new auditor, C. L. Hunt, a public accountant in Perth, was appointed as company auditor.

It must have been a bitter blow for the CSF to receive on 24 February 1987 the resignation of John Mackay, one of the original seven and acknowledged as a major driving force in the formation and ongoing success of the CSF. But his resignation after seven dedicated years rated no more than a single sentence in the broadsheet sent to subscribers. A further indication that Mackay's parting from the CSF may not have been all that amicable is that on the day he resigned he went into business for himself.

Corporate Affairs Commission records indicate the registration of a company, Creation Research Center, in the names of John Barry Mackay and Anne Heather Mackay, which commenced business on 24 February 1987. The nature of the new business is "lecturing on creation research and sale of associated materials." Home and business addresses are the same.

By March 1987, following the loss of more than $92,000 in investments and the various resignations, the CSF was left with only two effective directors, Osgood and Rendle-Short, both medical practitioners. Ken Ham, the third director, was resident in California, and has since resigned from the board. On 2 April 1987 new directors were admitted: Gregory Ross Peacock, described as a manager, and Charles Vincent Taylor, an educator.

On 7 October 1987 Bardsley was "removed" and new director Peacock appointed as company secretary. A few weeks prior to this, on 16 September 1987, the president of the CSF, Dr. Carl Wieland, was elected member, director and executive officer of the company. His home address is noted as Kewarra Beach, Far North Queensland, which is 1930 kilometers (over 1200 miles) north of Brisbane. Wieland is now the managing director of the CSF. Four months prior to Wieland's election, on 15 May 1987, Andrew Snelling, geologist and writer of articles for the *Ex Nihilo Technical Journal*, was elected member and director.

By 1988 the CSF had returned to the situation of seven director-members. One director acts as company secretary (Peacock) and another director (Wieland), presumably shifted down from northern Queensland, is both executive officer and managing director.

During 1988 the power structure of the CSF became more apparent. Wieland, Snelling and Taylor, together with Mrs. V. Wieland, were scheduled as speakers at seven seminars held throughout the country between April and November. Ham was overseas. Peacock, appointed three weeks after Wieland's election and on the same day that Bardsley was "removed," continued as company secretary while also fulfilling his main occupation as manager. Osgood and Rendle-Short continued as directors while also maintaining their careers as medical practitioners.

Thus Wieland, Snelling and Taylor were the three directors working full time with the CSF. Of these three, Wieland as marketing manager and executive officer is the key person in the organization. The more aggressive marketing strategies in 1988 also indicate this.

The two-year turmoil following the massive investment losses seemed to have subsided. The question to be asked is, what reassurance is there that the same thing will not happen again?

Although he wrote in 1987, Martin Bridgstock points to the problem inherent in the "closed shop" directorate of the CSF:

Far from being a broad-based movement with mass membership, the Creation Science Foundation consists of seven members who regularly re-elect themselves to the Directorships and who, one assumes, report to themselves about how things are going at the Annual General Meetings.

From the articles of association, it is clear that nobody can become a member of the Foundation without approval and so they remain as a small group in complete control of the Foundation's affairs.

The Foundation's current chairman, Emeritus Professor Rendle-Short, has attempted to answer these points. Writing in the evangelical magazine *New Life* (Rendle-Short 1986) he claims that:

"The Foundation is not a membership organization."

Exactly how an incorporated organization with a stipulated membership of up to 100 and an actual membership of seven is "not a membership organization" is rather unclear. Rendle-Short goes on to claim that the large number of people (about 12,000) on the Foundation's subscription and mailing lists "could loosely be described as our 'membership'." The looseness is extreme, and legally wrong. Members would be entitled to vote at general meetings, to demand a full explanation of the finances ... and, if the explanations were inadequate, to replace the directors. To our knowledge the subscribers and supporters were not told about the Foundation's financial problems until the first edition of this book brought the subject into the open. (Bridgstock, 1987, p. 82)

Finances of the CSF

There is no indication that the game of musical chairs which took place among the directors had any effect on the continuing growth of the CSF. Donations are a major source of income, as well as indicating the growth of the CSF. Donations grew from $33,000 in 1981 to $153,000 in 1984. Donations are not listed as a separate item after 1984.

Reports for 1985 were prepared by different auditors, following the resignation of the original auditors one week before the end of the 1985 financial year. Reports for 1986 and 1987 do not seem to be available. Presumably extensions have been granted by the Corporate Affairs Commission because of extenuating circumstances.

However, the statement of sources and applications of company funds for the year ending 31 March 1984 show that income from all sources increased from $316,281 in 1983 to $494,084 in 1984, an increase of 56 percent. Also between 1983 and 1984, bookshop sales increased 56 percent to $220,552 while donations increased 28 percent to $152,398. Even if total income increased at an ultraconservative 20 percent per year the 1988 income from all sources would be in excess of one million dollars. At 40 percent average annual increase, the total income for 1988 would be $1.9 million.

Whatever the actual amount is, there is in excess of one million dollars per year in the hands of seven men who between them fill the roles of members, directors, company secretary and executive officer/marketing manager/president. In addition, these seven provide no financial details to the thousands of "loose members" who donate so generously to the cause. They have revealed "extraordinary losses" only under duress, and the name of one of their directors, Robert Stephen Gustafson, who was also secretary, disappeared without explanation from company records after a payment of $8719 was made by the board of directors to a company in which he had an interest. He still serves as attorney to the company.

The March 1985 balance sheet shows that unsecured interest-free loans increased from $34,550 in 1984 to $51,050 in 1985. Presumably these figures represent the result of appeals for interest-free loans which appear in most issues of the CSF's range of publications. One wonders how secure these generous lenders would feel if they were made aware of the past performances of the seven directors and the musical chairs in which they seemingly indulge.

After the losses in 1983 and 1984, only $2 was left in the investment account. In 1985 this had increased to $25,000. Hopefully different investment advice has been taken. Since the latest records available are 1985 it is not possible to know whether the overall financial scene improved or worsened in recent years.

Loans and Donations

It is clear that from the very beginnings of the CSF donations have meant the difference between solvency and insolvency. A "Support

Form" which appears on the back page of the bimonthly broadsheet *Prayer News* calls for (i) donations, (ii) interest-free loans (minimum $1000) and/or (iii) regular monthly contributions. Potential donors are told that they can have donations paid straight into a creation science bank account, thus avoiding hunting around for money orders or making costly bank transactions.

The 1985 balance sheet indicates the trading surplus as $138,913 (mainly from book sales, video hire, etc.) while income from other items is $303,315 (mainly from donations). This gives a total income of $442,428 in 1985, but against this are overheads of $396,294 and an accumulated debt of $25,554. Thus donations are critical to the continuance of the CSF. The new marketing manager, Dr. Carl Wieland, is attempting an innovative strategy to increase book sales. Previously most sales were direct or through Christian bookstores and a few Catholic bookstores, although most Catholic bookstores do not stock creationist literature.

Materials Available From the CSF

There is a vast range of audio and video tapes available, along with numerous printed materials. The list is ever growing.

By the end of 1987 the number of audio tapes available was nearly 200, under the categories of archeology, biblical, education, family, history, language, law and government, science for the layman, science (technical) and miscellaneous. A large number of these tapes are talks given at public meetings and church halls around Australia. Only a few nowadays are from American sources. This is in contrast to the expensive videos and the many books, which are mostly from the United States, and the sales of which are major sources of income both by direct marketing and through the many Christian bookstores around the country. Lectures and seminars also provide a point of sale.

The journal *Creation ex Nihilo* is published four times a year and is a glossy color booklet containing the latest creationist discoveries. Some readers may find it enlightening to know that *ex nihilo* is Latin for "out of nothing." Most articles comprise either "scientific proof" that the world is only 6000 years old proving that the Bible (creationist version) is right or claims to disprove evolution in one way or another. The articles are usually interspersed with religious reflections.

Creation "out of nothing" is as mythical as Eve eating an apple. Neither appears in the Bible. Genesis 1 begins by stating that *before* God began His act of creation "the earth was formless and desolate. The raging ocean that covered everything was engulfed in total darkness" (Genesis 1:1 - 2). The biblical writers thought in concrete terms, not abstractions; they had no concept of "nothing." It is ironic that creationists can't even get right the very first verse of the Bible. The Semitic mind saw creation as "out of chaos" and the end of the world as falling back into chaos.

Ex Nihilo Technical Journal has appeared only twice, the first volume in 1984 and the second in 1986. It is a glossy journal which presents the latest in creation science while emulating in all respects a conventional scientific journal. It is for creation scientists the only vehicle for publication of their "scientific" writings on creationism. The second volume lists papers in the areas of biology, biblical history and archeology, geology, physics and astronomy and the speed of light. To anyone without current science expertise in these areas, the papers seem very convincing indeed. John G. Leslie's 36-page paper "Mutations and Design in Cellular Metabolism" has 167 references and is written at the level of a university textbook in biology. It concludes that mutations refute evolution.

The purpose of the technical journals appears to be to wave them in front of a gullible audience as evidence that creationists are "real" scientists doing "real" research. Another purpose may be to provide an outlet for creationist writers with some scientific expertise. None of the articles stands up to detailed analysis; nor have any undergone peer review by scientists.

Prayer News accompanies each of the quarterly issues of *Creation ex Nihilo,* which subscribers receive through the mail. Enclosed also is sales information on new books and videos as well as information on seminars, appeals for funds and a new car, and the latest research news on dinosaurs.

The August 1987 issue of *Prayer News* contains a one-page

rebuttal of Professor Ian Plimer, head of the geology department, University of Newcastle, who had strenuously attacked creation science in an article published in the journal of the Geological Society of Australia. At the end of the rebuttal the reader is asked to pray that Plimer's views will not prevail.

Prayer News contains information on how ministers can arrange for film nights and creationist speakers in their churches. Volunteers are requested to co-ordinate mini-conferences in their areas.

In 1988 there was a major change in the materials available from the CSF. In previous years books by American creationists were played down in order to enhance the image of an all-Australian enterprise. The president has publicly rejected accusations that the CSF was U.S.-derived. But this pretense has been cast aside in the need to implement stronger marketing strategies to increase sales.

A major item in the bundle of materials sent to interested inquirers in 1988 was the 32-page *Illustrated Book Catalogue*. This is a quick photocopy reprint of the total range of books and items available from the Institute for Creation Research in San Diego. There are in excess of 150 titles, under 14 headings, by 46 authors. The lone Australian author is Barry Setterfield. His book has been rubbished, his mathematics shown as ludicrous, his theory as ridiculous, and in addition he has been disowned publicly by Duane Gish. But still the book sells. The catalogue also lists Gish's booklet *Have You Been Brainwashed?* even though disowned by Gish himself with a total of 87 errors and "a lie every 11 words." And that one still sells, too.

Dinosaurs, Those Terrible Lizards by Gish and *Dry Bones ... and Other Fossils* by Parker are highly recommended as a "Children's Pack" so that children may have "in-depth materials." These are the two books which figured prominently in the notorious case at Smith school in Livermore, California, the two books which played a major part in causing gifted children to effectively vote atheist. Another title, *Origins: Two Model Series*, is available as a public school set or a Christian school set.

There are transparencies and texts for schools in physics, life sciences, biology, English, history of the world and cultures – all from a "Christian" perspective.

In all, this catalogue lists in the one place and for the first time in Australia the currently available books by any creation science author of note. It is a total spectrum of creation science literature and illustrates vividly that even when creationists are forced by public pressure to admit errors they just keep on publishing anyway.

Why the sudden change in policy? Why are American publications now given such prominence whereas previously they were billed as low key? There may be several reasons.

One reason may be that in 1985 overheads exceeded income from sales by $157,000. This means that if overheads since 1985 have increased faster than sales, the position could be reached where donations may be insufficient to cover the gap. This could be exacerbated if John Mackay's company, Creation Research Centre, which is a rival marketing organization for the same materials, starts to make major inroads into the CSF's traditional market.

A further major marketing thrust is pushing the sales of the nine-part video series *Understanding Genesis*. This has been produced at the Institute for Creation Research in San Diego. Four parts are by Australian founding director Ken Ham, and five parts are by prominent American creation scientist Gary Parker. The videos were introduced at most seminars, lectures and visits in 1988 and local groups were encouraged to show them at meetings. This series is marketed only in video format at a cost of $275 for the set. While profits from the sales of other materials go to the CSF, and employees receive a salary plus expenses, with this series Ken Ham joined the big league money-spinners in creationism. Royalties from the sales of creationist materials go to the author or producer and the world market for the *Understanding Genesis* series has enormous potential. Ham fared better than another founding director of the CSF, John Mackay, who apparently had to resign in order to achieve a just financial reward for his efforts.

The nine 16-mm films available from the CSF are another important resource. These are for hire only and intended for use in churches, seminars and schools. Six of these nine films comprise the *Origins* series. The first film in this series in the one in which Isaac Asimov appears without acknowledgment in the credits and in his own words "out of context." The hiring of the complete *Origins* series is $240 plus handling for a single showing.

All in all, the materials available from the CSF cater to every possible niche in the market, from the cradle to the grave as it were. More are in production.

The total number of errors, misquotes and deceits in this mountain of material would be impossible to count, even if one wanted to undertake such an effort. It is understandable only in the context of a market ever eager to maintain a fragile religious faith which would shatter if evolution were admitted into it.

Marketing Techniques of the CSF

The CSF describes itself as "Suppliers of quality Creation Science material, Christian School texts and Audio Visuals." It sells directly and through bookstores. Materials are also on sale at all meetings, lectures and debates.

There are agents in every state. In the past decade networks of willing entrepreneurs of creation science have been constructed throughout the cities and country towns of Australia. Through their nationwide seminars the three active directors – Wieland, Snelling and Taylor – are seeking to expand the number of people prepared to return to their local churches, fellowship groups and schools, there to introduce creation science as well as run their own seminars using the wide range of materials available in audio, visual and printed form. Correspondence courses and training tapes are available for the serious student of creation science. In addition, there is support from those in authority. Creation science is often welcomed as certain priests, ministers and lay leaders seek to verify their own fundamentalist beliefs in front of their flocks.

It is likely that most of the young people first exposed to these materials are literalist in their biblical beliefs and mildly fundamentalist in their doctrinal beliefs. But exposure to creation science materials changes all this.

Evangelical organizations such as the National Alliance for Christian Leadership in Canberra are willing partners in the spread of the creationist network. Perhaps for some, such activity that results in warm acceptance by their flocks gives a sense of achievement as well as a sense of power. They are the leaders who have sponsored the prophets from afar. Some may believe in what they are doing. Others justify lying when it is done in God's name. A third group may undergo the metamorphosis from lying to hypocrisy. A hypocrite is a person who finally believes his own lies.

In Australia, as in the United States, such single-mindedness, tenacity and sheer determination of but a handful of individuals would be enough to create a major wave through society.

New Star Is Born

On 24 February 1987 John Barry Mackay decided to act. He resigned from the Creation Science Foundation and immediately set up his own business in partnership with his wife, Anne Heather Mackay. John Mackay has been prospering mightily since that day in February. Twelve months later, Mackay's center boasted a staff of five, not including himself. He had all the trappings deemed proper for a creationist enterprise – a Statement of Faith, "ministry" around the country, a catalogue of books and tapes, pleas for donations, and so on. In an obvious swipe at the CSF's missing $92,000, all funds lodged with Mackay's CRC are guaranteed.

There is no pretense that the CRC is either a science organization or a nonprofit-making foundation. The *aims* include researching and documenting historic and scientific evidence for Christ, creation and the biblical record, proclaiming the Gospel in schools and tertiary institutions and, finally, encouraging Christians "to win others to Christ by motivating them and by making available for purchase, the materials to do it with at the lowest possible cost" (*Who? Why? What?*, a single-page undated broadsheet obtainable from the CRC on application).

The Statement of Faith on this broadsheet is divided into four parts. Part A is similar to the CSF aims, stating that the doctrines of Creator and creation cannot be divorced from the Gospels. Part B emphasizes the inerrancy of Scripture, repeating Morris's intriguing statement that "the final guide to the interpretation of Scripture is Scripture itself," as well as emphasizing the historical truth of creation in six days, the Noachian Flood, death entering the world through Adam's sin, and so on. Part C is reminiscent of Christianity in bygone ages. Mixed up with creeds and bits of doctrines on original sin are such statements as "Those who do not believe in Jesus Christ are subject to everlasting conscious sin but the believers receive life with God." Part D contains two statements which reflect Mackay's personal belief: "The common view of knowledge into 'secular' and religious is a false view" and "The chronology of secular world history must conform to that of biblical world history."

As befits a school teacher turned minister, the catalogue of books and audio tapes emphasizes the needs of children and youth. Mackay is energetic to a high degree in the pursuit of his goals at the grassroots level. In the month of May 1988, he completed four weeks of ministry in Adelaide, Melbourne and Sydney. He spoke to science teachers from four high schools in South Australia, and in thirteen schools he addressed several thousand students. In the same four weeks he spoke "at approximately 30 churches and public meetings" (*Creation News*, CRC, June 1988). In this newsletter Mackay describes his meeting with the new Queensland Minister for Education, Mr. Littleproud. He tells his readers:

> You'll be pleased to know (Mr. Littleproud) assured us he intends to continue the previous Minister's (Mr. Powell) open-handed policy on the teaching of creation in Queensland schools. Which means for the present time, teachers in Queensland are free to comment on creation without fear of persecution.
>
> Unknown to most people, the Director General of Queensland State Education is a Catholic, the Minister of Education attends a Catholic church and the Premier is also a Catholic. It is interesting then, following the opposition of much of the Catholic Education Department to teaching creation, that the Queensland Education Department has now approved the tape series "Evolution Takes on God" by Wallace Johnson (also Catholic) for use as resource material in schools and school libraries. These tapes which use

material essentially derived from Dr. Parker and other creation speakers are yet another indicator of the Government's willingness to be open-handed on the creation issue as opposed to their southern counterparts who are violently anti-creation in so many cases. (*Creation News,* CRC, June 1988)

On the same page readers are asked to pray for John while he spends two weeks in Sydney:

As many of you know, the New South Wales government has told Christian schools they will be shut down if they teach creation in science classrooms. Satan is really rearing his head through the government of New South Wales and we have no option but to speak and write against such things. There is an election in New South Wales in March and ALL Christians must stand and be counted and we will be encouraging them to do just that! Men and women of God everywhere must raise up and declare the mighty works of our God before the believer and unbeliever alike. Join us as we stand at the battle front!!!

Since the above newsletter the government was defeated. But it seems that Satan still holds the reins in New South Wales. The new government has made no moves to rescind the policy of the previous government. No doubt the Minister for Education is being bombarded daily with letters from creationists.

When he was a member of the CSF, now his rival, John Mackay wrote articles jointly with Andrew Snelling. Both ardently believe that fossils can be formed very quickly. In June 1988 Mackay returned to the Grand Canyon where on a previous trip with the ICR's Gary Parker, he found "large amphibian-like footprints." As the June 1988 newsletter notes:

It has always been our practice to schedule research between ministry engagements. That way it involves no extra travel costs. In June/July John Mackay will be researching everywhere from the coal mines in British

Columbia, through the Grand Canyon, across to the polystrate trees in Nova Scotia and through the swamps of Florida. All this is fitted in between preaching and makes for an exhausting but profitable trip.

Mackay is very much a faithful follower of Henry Morris. However, Mackay is much more than a younger and more energetic version of his CSF counterpart, Dr. Carl Wieland. The notes and worksheets he has written for the training of high school teachers in the execution of creationism in the classroom testify to this.

Same Tune, Different Key

Brisbane Grammar is a very prestigious, and expensive, private school in Brisbane. On 15 August 1988 it was host and venue for a course by John Mackay on the teaching of creationism in schools. The participants were high school teachers from Brisbane Grammar itself and other independent schools, that is, schools which are neither public nor Catholic. So-called independent schools usually have a religious tradition and the title of Grammar applied to the older establishments usually indicates foundation by the Church of England.

The handouts at the course included a list of resources for the teacher and the classroom. No matter the level of sophistication of the worksheets and course materials, the resources listed indicate without any shadow of a doubt that it is the same old theme with a few new variations. All the materials are stocked and sold by the CRC, Mackay's business. By now you will be familiar with some of these. The four children's books listed include *Dry Bones ... and Other Fossils* by Gary Parker, a personal friend of Mackay. This is the same notorious book used at Livermore whose function is to inculcate that creationists believe in God while evolutionists are atheists.

Because participants at the course were science teachers the list included thirteen technical monographs, with titles relating to the origin of life, radiometric dating, the age of the solar system, the age of the earth, the origin of the universe, the age of the cosmos and catastrophes in the earth's history. In sum, all the items necessary to convince the layman, and many science teachers, of a 6000-year-old universe, the Great Flood, etc. etc. These technical reports are written with copious references, to look like genuine scientific papers. Except that no reputable scientific journal would publish them. They don't stand up to expert analysis.

There are six textbooks, for the library or the teacher, which give world history, geography and world cultures from a "Christian" perspective. This means that all humans descended from the survivors of the Flood and so on. Shades of post-millennialism. There are a dozen or more children's books ranging from nursery to kindergarten through to *Biology: God's Living Creation*, costing more than $30.

The pièce de résistance of Mackay's selective list is the twenty-two creationist books for general reading and the library. These range in price from a dollar to over $35. Five of the books are by Henry Morris, including *The Genesis Flood* (Whitcomb and Morris). Books by Parker, Bliss and Wilder-Smith also rate a mention. In sum, these books comprise a compendium of provenly effective creationist publications. The selection includes books which, taken together, comprise all creationist arguments for creation according to the Bible.

Last but not least on the list is a set of four tapes called *Creation Information Teaching Seminar*. These enable science teachers to pass on the good news of creationism to their colleagues.

In all, the comprehensive package recommended by Mackay is guaranteed to make any science teacher into an expert. There is one gap, however. There are no books specifically on Satan in this list of books recommended for schools, unlike the general catalogue where one example is a tape, *Turmoil in the Toy Box*, which explains how Satan is "after your kids through their toys."

The Best Is Yet to Come

The book catalogue handed out by Mackay at the Brisbane Grammar School conference of science teachers shows, to those familiar with the texts, that the Creation Research Center is promulgating the same creationist line as the grandparent ICR in the United States and the parent CSF in Queensland. However, the papers handed out at the inservice conference for science teachers indicate that the child has reached a level of sophistication in the dissemination of creationism which surpasses both parent and grandparent.

Four of these will be described briefly. Two of them are unsigned photocopied documents which summarize strategies for implementation as well as strategies for the teaching of origins. The second two are apparently draft chapters of a book; each is about 3000 words and marked "copyright J. Mackay 1988."

In the unsigned handout called *Implementing the Teaching of Creation and Evolution* point one suggests that

A lot of the emotionalism can be removed from the whole origins debate by increasing student awareness of the limitations of science particularly in respect to historical events and the origin of life. It would be of value for students to be aware of the biases of scientists and how these do color their search for evidence and their conclusions, particularly in the section dealing with the origin of life.

This sounds admirable but suggests that the millions of scientists from a hundred nations and a dozen religions, or no religion at all, have been acting out of religious bias in concluding that evolution is fact. If one asks where science exists, it is in the millions of peer-reviewed scientific journals spread throughout the libraries and institutes of the world and not in the mind of a single scientist. Scientists and mathematicians like all humans suffer from bias and prejudice. But if this creeps into the writing of their professional results they will soon be looking for another job. Science and mathematics, insofar as they *are* entities, are like nature itself, which is neither moral nor immoral, biased or unbiased. It just is.

The document continues:

It is also an excellent area to consider the social implications of scientific endeavor. This is particularly true since every concept of where man came from has an inherent value system associated with it.

It is true that there is nowadays a very much increased awareness of the social implications of science and, even more so, of the social responsibility of science within the scientific endeavor itself. But science operates within a culture and it is the culture or society which puts values on science. This is most evident when governments fund particular areas of science. If Mackay wants to look at value systems attached to theories of origins then he should examine some of the hundreds of creation stories from world religions. The ICR's Statement of Faith gives the definition of creation as that found in the Bible. Any comparison with other religions would show clearly that his whole scientific edifice is based on just one of many ancient stories.

In effect, what Mackay is doing is a more sophisticated version of what Baird did at Livermore. He sets up a basic premise that creation and evolution are both just theories and hence of equal validity, but in a more sophisticated way than Baird. The rest of the document describes the particular needs of the junior and senior high school, inservice training, materials development and resource materials. He comments on the need to make librarians aware that creationist materials should be catalogued under science and not religion as many librarians are wont to do.

A second handout, marked "draft," is called *Teaching Origins and Development of Life*. This little paper, about a thousand words, displays a palpable ignorance of modern Darwinian theory. This is not unexpected. As at Livermore there are no competing textbooks. One gets the impression that books such as the 1986 classic best-seller on this topic, *The Blind Watchmaker* by Richard Dawkins, are on some secret Index of Forbidden Books.

The handout distinguishes between "the development of" and "the origin of" life but falls far short of making the distinction between evolution and theories of the origin of life. The two are quite different. Evolution is a fact supported by evidence from many branches of science, while the way in which life first occurred is still under investigation. There are several competing theories of origins.

Perhaps the gem of this handout is the list of eleven scientists who believed in creation. Among these we find Francis Bacon who wrote in the sixteenth century and was savage in his criticisms of biblical literalists; Isaac Newton, who lived long before Darwin; Werner von Braun, the German-born rocket scientist, is the only one alive today and hardly an apt example. Most of the rest lived before or about the time of Darwin. A startling omission is failure to mention the many thousands of scientists who believe in both God and evolution. These far outnumber the sum total of the relatively few scientists who claim to be creationists! The main thrust of the document is repetition of the implied assertion from the first paper that scientific theories are based on *religious* values. One wonders, what religion is Ohm's law in electricity based on?

The document *Detecting Design* is copyrighted to J. Mackay and labeled "textbook, Part 5." It repeats in the form of worksheet notes what is essentially Gish's argument that the laws of probability rule out evolution. This is a very convincing argument to those ignorant of modern science. It is culpable to teach this to children as representing the views of science today on the mechanism of evolution. This whole argument is disposed of in Chapter 11 of this book.

The fourth handout to be discussed here is called *Unit 10, Philosophy/Religion and the History of Life*. It attempts to show that a belief in the way the world came to be determines belief or rejection of God. The concepts of origin treated are chance evolution, initiated evolution, controlled evolution, creation-limited variation and creation without variation. It is more than a little ironic that this last-named group is labeled as having a basic common factor of "apparently rigid conservatism or opposition to change, biological included." Apparently this refers to the ultraconservative branch of creationism. The worksheet is the fill-in type which gradually leads to the acceptance of the right answer by a pseudo-objective line of reasoning.

It demonstrates, if nothing else, a total misunderstanding of the Christian doctrine of creation. This is entirely independent of *how* the world was made. The *doctrine* sees creation as a relationship between created and Creator. That God made the world in six days is only a theory of the origin of the world.

The CRC in Summary

The leopard has not changed its spots. If, as the old saying goes, the greatest error is the one closest to the truth, then the CRC is a supernova among the stars of creationism.

In conclusion, one can only wonder if the parents who send their children to Brisbane Grammar and other independent schools are aware of what is going on. A large proportion of parents are hardheaded business people, wealthy industrialists, graziers and the like. It is difficult to believe that they would all agree with what is being taught to their children. Likewise, the school boards to which the principal is responsible surely can't be made up entirely of creationists.

Science Strikes Back

Henry Morris's Institute for Creation Research cloned itself into Australia in 1980. The Creation Science Foundation found fertile fields in Australia. By 1984 the creation-evolution debate surfaced in a number of professional journals. Public debates on the issue took place and the topic began to appear in the media.

Some Education Departments took notice. As well, research which showed that a significant number of students entering biology at university level did not believe in evolution acted as a spur which caused more scientists to become involved. Science writer Bob Beale (1988, p. 74) says about the Gish/Plimer debate:

> Observers in the science community see that recent debate as a turning point – that an anti-science battle being waged on many fronts is pushing more and more scientists into fighting fire with fire.

Beale describes Plimer's tactics and the aftermath:

> Dr Duane T. Gish, a leader of the US creation science movement, will never forget the night in Sydney when he took on Professor Ian Plimer in an extraordinary debate about the theory of evolution.
> For more than 20 blistering minutes, Professor Plimer mocked, ridiculed, and challenged every tenet the movement holds dear, and made a string of blunt personal allegations about some of its more prominent members.
> At one point he even donned insulating gloves, took a live electric wire and offered Dr Gish the opportunity to electrocute himself.

His point was that creationists would selectively accept that the science of electricity could be based on theory, but not the science of evolution.

A visibly moved Dr Gish accused him of being theatrical, abusive, mudraking and slanderous. "May I say it was the most disgusting performance I've ever witnessed in my life," he said.

Professor Plimer, who heads Newcastle University's geology department, is worried that the creationist wave is part of a broader backlash against the information boom.

His response has been to go in boots and all, aiming for his opponents' kneecaps, exemplified by the fact that much of what he said in the Gish debate cannot be repeated for legal reasons.

He is emerging as one of the toughest opponents creationists have encountered anywhere, using tactics he says he learned in the mining world.

"I take the Broken Hill approach: you don't put up with bullshit, and you take no prisoners," he said yesterday.

"I'm doing what other scientific people have never done, and take them on in their own way.

"Essentially I regard it as a political exercise. I'm not going to argue about spots on butterflies or the speed of light.

"You can't argue science with someone who wants to promote religion as part of a school science course: it's like arguing that equal time be given to witch-doctoring in a medical course."

For his pains he is facing several defamation threats, is the subject of letter campaigns, and creationist leaders are pressuring Newcastle University to discipline him and publicly apologize to them.

Few scientists, however, yet share his taste for confrontation. (Beale, 1988, p. 74)

However, while Plimer's approach is the most visible, it is underpinned by the works of many other scientists in providing the research and thereby many of the bullets, which are essential for confrontation against creation science on a wide front. These are necessary if the community at large is to gain an understanding of the science issues involved. They also illustrate the increasing number of Australian scientists who have come to see the issue as important.

Confronting Creationism: Defending Darwin is published by the New South Wales University Press in conjunction with the Australian Institute of Geology. This book gives the lie to almost every bit of pseudo-science used by creationists but goes much further. It also confronts each error with an accurate picture of the current science. It is an invaluable resource for science teachers and for anyone interested in science. In *Creationism: An Australian Perspective*, published by the Australian Skeptics, Martin Bridgstock (1987) gives an overview of creationism in Australia. Since 1986 it has been kept up-to-date and July 1987 marked the third (revised) edition. It covers a spectrum of creationist errors in brief and several in depth as well as providing examples of misquotes and an analysis of the structure, aims and activities of the Creation Science Foundation.

In their attempts to suppress evolution, creationists complain to radio and television stations when programs mention the age of the universe or stars as being millions of light-years away. In 1986 creationists threatened the Commonwealth Bank in order to secure equal time in the Australian Museum's traveling train exhibition, What's on Earth:

> The museum train travels throughout NSW and is visited by about 100,000 people a year, mainly schoolchildren. It was sponsored for $50,000 last year by the Commonwealth Bank.
> The sources say creationists have subjected the bank to a letter-writing campaign for more than a year, objecting particularly to one paragraph of the text displayed in a section of an exhibit dealing with evolution.
> The letter-writers, including a major Western Australian investor, threatened to withdraw their deposits from their accounts with the bank if the offending section was not altered.
> "I know they (the bank) were frightened enough to ask us very strongly to change that particular paragraph," said a museum spokesman.
> The museum decided to dig in its heels and altered the text

to state even more strongly that of all the theories about the origins of life, the theory of evolution was by far the one given most credence by scientists, the spokesman said. The bank had been happy with the change.
A Commonwealth Bank spokesman ... said the bank sponsored the museum train on educational grounds, and had every confidence in the museum. The bank did not seek to exercise day-to-day control over any of its sponsorships. (Quoted from *Sydney Morning Herald,* 15 February 1986, in *Creation/Evolution,* vol. 6, no. 3, p. 9)

However, in September 1988, the Australian Museum replied in spades to creationist critics. The museum opened a spectacular exhibition on human evolution, called Tracks Through Time. Of special note and a world first in any exhibition of this type are the four words which greeted visitors as they entered the exhibition: Evolution is a fact. The exhibition was opened by Dr. Richard Leakey, famed for his studies in Africa on the origins of man.

Not Just Science

It should not be taken from what has been said above that Australian scientists see creationism as merely a threat to science. Dr. A. G. Wheeler is a scientist at the University of Queensland and author of *The Other Quote Book: In Support of Evolution.* In an article, "Creation Science in Australia," published in *Quadrant,* he sees creationism as a threat to a free society. He quotes Isaac Asimov in support of this:

Isaac Asimov sees the threat of creation science in these terms:
"Today 'equal time', tomorrow the world. Today it is your views on science, tomorrow the way you dress and speak and behave. It is not merely creationism that we are fighting in this matter. Behind it are the old enemies of bigotry and darkness, and we must not complain about this endless battle. The price of liberty, said Jefferson, is eternal vigilance."

Science, education and individual liberties are the main targets of creation science and their victories will be our losses. (Wheeler, 1987, p. 61)

In the conclusion to the same article Wheeler gives his views on what creation science is really about:

If you examine the objectives of the CSF, their methods, practices and publications, you will be impressed by the relative paucity of religion. The CSF does not appear to be in the business of saving souls, converting sinners or enhancing man's relationship with God or offering salvation through Christianity. Rather their business is counting the number of invitations to speak against evolution, urging librarians to stock their anti-evolution material, bringing pressure to bear on librarians, teachers and others who dissent from their views, and gaining political influence to further their own ends.
In the opinion of Ronald Strahan, editor of *Search*, the journal of the Australia and New Zealand Association for the Advancement of Science:
"Fundamentalism can be tolerated while it remains restricted to self-centered, exclusive minorities but it endangers the fabric of society when it achieves power."
On the surface the CSF is a religion masquerading as a science. The CSF represents an anti-liberal, pseudo-science, pseudo-religion that aspires to political power to censor our morals. This, believe it or not, is what the debate over evolution is all about. (Wheeler, 1987, p. 62)

Chapter 15

Churches and Causes

Across the spectrum of creationism God comes through as angry and vengeful, one who sends millions of men and women to eternal damnation. This is the god of literal Genesis, the god whom evolution would drive away but is not allowed to because so many can see no other god. This god is at the very root of creationism. Understanding how this god was made is essential to an understanding of the hold over the human mind exerted by creationism. But this god is not a new god. He was not invented by creationists.

Theologian Christian Duquoc quotes Friedrich Nietzche (1844 - 1900) who was writing about the Christian God when he said:

> The ancient biblical legend believes that man possesses knowledge. His expulsion from Paradise is simply the consequence of this; from then on, God fears man and immediately expels him from the place where the tree of life, immortality, is to be found. If man was now also to eat of the tree of life, his power would have been established. All civilization would enhance the impressive character of man, as symbolized by the Tower of Babel, with its attempt to "capture the heavens". But God separates men, and scatters them; the plurality of languages responds to his urgent need; God is far better off since from now on the various nations war on each other and destroy each other ...

At the beginning of the Old Testament we find the famous story of God's agony. Man is represented as God's error: work, necessity, and death are presented as God's "legitimate defense" established to keep man in a debased state ...
The agony of God: man as an error of God; the animals as well. From this one can draw the moral of the story:
"God forbids knowledge because it leads to power, to divinity. He would have consented to man's immortality in itself, provided that man remained immortally stupid.
"God created animals, and then woman, in order that man have company and entertainment (so that he would not have wicked thoughts, and begin to think and to know)."
But a mischief-maker arrives – the demon, symbolized by the serpent, who reveals to man the final word on the situation:
"God's danger is enormous. Now it is necessary for him to drive men far from the trees of life and hold them down by distress, death, and labor. Real life is represented as a legitimate defense of God, a state against nature ... Nevertheless, civilization, the work of knowledge, aspires to equality with God; it rises up like a tower in an attack on heaven. From then on war is necessary – men should be brought to destroy themselves. Finally, the end of the world is decreed.
"And it is in such a God that people have believed!" (Quoted in Duquoc, 1978 - 79, pp. 193 - 4)

The Jews, from whom Christians borrowed the first part of the Bible and renamed it the Old Testament, have never interpreted Genesis in the Christian way. Nor have they even attempted to pinpoint the cause of evil in the world. The Eastern mind is more prone to let understanding stop at what cannot be understood. The book of Job examined the problem set by the author of Genesis who proclaimed that everything, both good and evil, was made by God. His emphatic conclusion was that the solution was beyond man's understanding. Yet for 1500 years the description of the god described by Nietzche was embedded into the culture of the

Christian West, albeit balanced by Christ and the prophets. The departure of this god from the scene is still unknown to many. The fault surely is with the leadership of the Christian churches. Why the silence? A silence which must bear part of the blame for the rise of creationism.

The Best-Kept Secret

The year 1872 witnessed the archaeological discovery of the tablets which told the source of the Genesis myths of creation. By the 1950s at the latest, there was agreement among biblical scholars that the first eleven chapters of Genesis were not historical. By the 1960s theologians were writing in terms of evolution. Today, theologians accept evolution as fact. In 1963 Catholic theologian Raymond J. Nogar could write:

> After one hundred years of careful research and enlightened discussion, scientific evolution is generally accepted as a fact by the educated person. The days of religious polemic among scientists, philosophers, and theologians are over. As the dust of former battles has settled, the field of controversy has been cleared by mutual understanding. (Nogar, 1963, p. 169)

The question which puzzles many is that a large proportion of the general population, both Christian and non-Christian, seems unaware that evolution and church are no longer in conflict. Protestant theologian Langdon Gilkey comments upon what he labels the best-kept secret.

> (It is) an error almost as bizarre as if most of the public thought that the majority of contemporary doctors *bled* their patients regularly as they were wont to do with deadly effect less than two centuries ago! (Gilkey, 1985, p. 187)

He adds that "the century-old rapprochement between science and theology is the best-kept secret in our cultural life." He also notes that a feeling of inadequacy in dealing with scientific matters may influence some pastors.

Especially we put it off when we fear that offended lay members may leave and join one of the conservative churches down the street. Thus, while "God's creation" may frequently be ringingly affirmed, a careful and explicit interpretation of creation as in accord with modern science rarely appears and the secret is kept well under cover.

In short, then, if the scientific community is partly to blame in this controversy for heedlessly assuming before the public that scientific theories simply replace "religious myths," so *the leadership of the churches is as much to blame for querulously avoiding an important issue and thereby keeping the actual fact of a theological aggiornamento (updating) quite hidden from view.* (Gilkey, 1985, p.128)

Problems in Rome

While Catholic theologians and Scripture scholars accept evolution as fact and assume it as such in their writings, there is an ambivalence on the issue that indicates some confusion at the highest levels of the Roman Catholic Church. This has resulted in many contemporary Catholics perceiving evolution as highly suspect and at worst heretical. It has also given Catholic fundamentalists a free rein.

Theoretical physicist Stephen Hawking relates how in 1981 he attended a conference on cosmology organized by the Jesuits in the Vatican. Hawking states that at the end of the conference Pope John Paul II told the participants that "it was all right to study the evolution of the universe after the big bang, but we should not enquire into the big bang itself because that was the moment of creation and therefore the work of God" (Hawking, 1988, p. 116).

This ambivalence is also illustrated in a more recent statement of Pope John Paul II:

It can therefore be said that, from the viewpoint of the doctrine of the faith, there are no difficulties in explaining the origin of man, in regard to the body, by means of the theory of evolution. (John Paul II, 1986)

Here Pope John Paul II goes much further than the statement of Pope Pius XII in his 1950 encyclical *Humani Generis*, wherein he sees extreme difficulty in reconciling Adam and Eve with polygenesis. This difficulty of Pope Pius has been resolved but there remains by implication the former mindset that creation as described in the Bible and evolution are both theories, on equal terms.

It may well be that Rome, like Gilkey's ministers, is afraid that the congregation will walk down the road to a more conservative church. On the other hand, it may be a surviving remnant of the mindset that evolution and church doctrines were irreconcilable. This was part and parcel of that reactionary phase of the Roman Catholic Church known as the "modernist crisis" in the late nineteenth and early twentieth centuries. Although no longer of consequence, it "spanned the entire theological careers of many twentieth-century scholars and the entire intellectual formation period of the overwhelming majority of priests ordained in this century." (McBrien, 1980, p. 54)

Whatever the reason may be, the ambivalence at the top level of leadership is a major factor in abetting Catholic fundamentalism. This is a powerful force and includes large numbers of priests, laity and some hierarchy. Another outcome is that fundamentalists use Church statements to label as heretics those who accept evolution. The same Church documents are used by others to show that the Church accepts evolution.

In all this turmoil and confusion there is a deafening silence from those at the top when Catholic fundamentalists ally themselves with creationists in their attempts, albeit sometimes successful, to have creationism taught in Catholic schools. No one can gauge the effects on their congregations – not just on those who become adherents of creationism but also on those believers in evolution who are driven away. Catholic fundamentalists persist in stating that Catholics cannot believe in evolution.

Are creationists entrenched in Rome? Or is there some other reason? Is there more to the issue than a reluctance to abandon direct creation in favor of evolution?

Specter of Original Sin

In 1950 Pope Pius XII felt that it was extremely difficult to accommodate the historical Adam and Eve within the framework of evolution. Before that, in the early decades of the twentieth century and during the modernist crisis, the Roman Church effectively ruled evolution impossible for the same reason. The naive creationist version of the second law of thermodynamics likewise seeks to protect the historicity of Adam and Eve, thereby preserving the ancient Genesis story that all humans are descended from a single pair of individuals who were instantly created in a state of perfection.

Creationists are locked into a position similar to that of the Catholic Church in the early twentieth century. The difference is that instead of using doctrines and superseded teachings to declare evolution impossible, they use their own version of the second law and the literal reading of Genesis.

The dogged defense of Adam and Eve as real people is because in the fifth century the great St. Augustine, under extreme pressure from the gnostics and in opposition to the Irish monk Pelagius, used a literal reading of Genesis to argue that evil entered a perfect world through the sin of disobedience by Adam and Eve. The gnostics were a powerful influence in the early centuries of Christianity. They believed in two gods or cosmic principles, one of evil and one of good.

Saint Augustine's solution, much simplified, was that the disobedience of Adam and Eve caused sin and evil to enter the world and that, moreover, this sin was inherited by their descendants, the whole human race. This became known as the Doctrine of Original Sin. It was ratified by subsequent councils of the Church and an important decree document by the Council of Trent (1545 - 63) was that on original sin. Even the position of *Humani Generis*, the 1950 encyclical of Pope Pius XII, concerning polygenism, is based on the decree of the Council of Trent (Alszeghy and Flick, 1967, p. 199). This interpretation of Genesis 2 and 3 "has exercised a considerable influence on the Western Church and our culture as a whole" (Duquoc, 1978 - 79, p. 189).

Telling the Story

Saint Augustine is justly regarded by many as the greatest of the Fathers of the Church. Much of his writing has a timeless quality which echoes down the centuries, finding a deep response in the hearts of men and women in every age. He taps the very depths of the human condition and his *Confessions* is a classic piece of literature in its own right. As is not uncommon with persons of greatness his work is sometimes misinterpreted and misunderstood by his followers. Because this section is a narrow focus for a specific purpose, space does not permit a full explanation of the constraints and pressures under which St. Augustine labored. Many modern writers have done this. One of these is Father Christian Duquoc, a French Dominican theologian, in his article "New Approaches to Original Sin," which first appeared in French in 1977.

Before noting some of the alternative interpretations by modern theologians in the latter part of his article, Duquoc states that "original sin has now become a scandal" (p. 190). He also states:

> The Augustinian interpretation of Genesis 2 - 3 became a dogmatic structure which no longer responded to any vital debate for Christianity; it became a gratuitous affirmation, burdening God with injustice and cruelty. It appears outrageous that for an archaic fault, whose content is unknown, men are made guilty before God and hence punished by all kinds of evils – a curse on the soil, the disordering of desire, the alienation of woman, the burden and brutal character of labor, and death itself. (Duquoc, 1978 - 79, p. 193)

Thus to admit evolution means not only abandoning six-day creation but also abandoning the 1500-year-old belief on how sin entered the world. Nor are alternative answers readily available. Under these conditions acceptance of evolution becomes a giant leap into the unknown beyond the capacity of many. But there is another consequence which makes it even more difficult for creationists and Christian fundamentalists. This is the answer to the next question asked by Augustine 1500 years ago: "Why did God become man?"

Answers Are Easy, the Questions Are Difficult

The 1200 years from the early Fathers to the Reformation saw Christianity concerned mainly with problems of mortality, death and preoccupation with sin, while salvation was seen in terms of eternal life. The Reformation period (sixteenth century) was preoccupied with guilt and wondering whether anyone was perfect enough to get to heaven. The doctrine and story line of original sin had seeped into every crevice. In contrast, the preoccupation of Christians in the early centuries was with how to get to heaven, while the attitude of the Old Testament was more thankfulness and praise to God for the gift of life. But "today's crisis in an age of historical consciousness, pluralism and relativity, is the question of meaning" (Haight, 1985, p. 37).

The story line of God making a stunningly beautiful creation where there is no pain, suffering or death and placing humans on it, with the foreknowledge that they would sin, and moreover, imposing a penalty of eternal damnation, is simply not relevant today. It lets God "off the hook" as it were by placing responsibility for the evil in the world on the shoulders of man but at the expense of creating a cruel and vindictive God.

Furthermore, the question was asked, "*Why did God become man?*" The inevitable answer was, to redeem humanity from original sin. Another effect of abandoning the Augustinian version of original sin has been to emphasize more strongly the role of Christ as the Savior of mankind rather than as Redeemer of a nonexistent sin.

This brings us to what may well be the crux of the whole issue. Virtually all reflections and considerations of the doctrine have been done in the mindset used by St. Augustine, which is: of an infinite number of possible worlds God chose to create this one – *the best of all possible worlds.* Nor is there any other possible conclusion in the now obsolete cosmology of a static unchanging universe. As shown in Chapter 11, if one were to ask the same questions as St. Augustine again today, it would be out of the mindset that to the best of our knowledge at this point of time, *this is the only way God could have made the world.*

If one were to continue, and to speculate further, on why there is both good and evil in the world – a mightily presumptuous question and one which only the Western mind would be so foolish as to ask – one would certainly come up with an answer very different from that which St. Augustine was forced to make. Using the images to which we are limited by our own humanity it is conceivable that St. Augustine, if writing today, may have envisioned a story such as the following:

A god of infinite power and love existing outside of space and time, in timeless time, wished to share his own essence, which is both to be and to love. There is an infinity of hesitation because the only possible way of creation brings into being suffering and death as well as joy and love. The hesitation is the same as parents deciding to bring a child into the world knowing that pain and suffering are the human condition. The decision is made with the same trusting hope of parents that life, love and joy in creation itself will outweigh pain, suffering and death. Evolution begins.

This is, if nothing else, consistent with John 3:16, "For God loved the world so much that he gave his only Son ..."

Such a story by St. Augustine would have given a different direction to the following 1500 years of Christianity. At present Christianity is without a credible theology of creation. The emerging creation theologies may bring us such a story. The need is urgent. The planet itself is endangered and the best-kept secrets may stay secrets until it is too late. At present Christian leaders seem blissfully unaware that the threat to our environment is such that preoccupation with nuances of doctrine, changing moral attitudes, and so on, may become irrelevant simply because there will be no one around to care about such things.

Ignoring the Needs

The churches today have responded to the needs of the world. Famine relief, social justice, the poor in the cities all are seen in the context of the parable of the Good Samaritan on a global scale. Moral arguments against nuclear arms indicate a willingness to proclaim Gospel values. But atheists, humanitarians, other religions, governments both democratic and socialist as well as the United Nations, are equally committed to these same issues in the name of humanity itself, in the name of human beings which existed before, and take precedence over, all religions.

But there is silence on the one in most need of all. The planet itself, the Earth Mother, is dying and the symptoms are there for the whole world to see. Yet from the Christian church leadership there is not a sound. It has yet to be considered a Gospel value.

The earth is still seen as a launching pad to heaven, a waiting room. Failure to read the signs of the times indicates a creationist mindset present at the highest levels.

Before pursuing this it must be said that many scholars and theologians working, as it will be shown, within strictures imposed from above have responded to the need. In the interim the battle to save the planet is being waged in the main by humanists, green activists and environmentalists. If the religious energies of creationists could somehow be reversed and pointed toward saving the planet, no power on earth could stop them. If not prevented, so would the energies of a billion or more Christians. Instead, such issues as whether Marxism is creeping into some theologies, whether masturbation is still a deadly sin, divorce, contraception, married priests and a host of other issues, some of them admittedly important, occupy and preoccupy those in top Roman positions.

But it is not too late to learn. Urgently needed is a theology of creation based on evolution and a corresponding creation spirituality. Spirituality simply means a way of relating to and experiencing the divine through the world around us. It means emphatically and publicly discarding *all* of St. Augustine's version of original sin. As will be shown, the groundwork already exists at the grass-roots level. It means that *Rome must embrace evolution as fact, not theory,* as does the rest of the educated world. It is ironic that the Church, with its own Pontifical Academy of Science and its own observatory which is regarded as at the forefront of investigations in cosmology, refuses to listen.

Emerging Solutions

Many theologians, both Catholic and Protestant, are engaged in the designing of living theologies of creation. Each single one of these uses evolution and our current understanding of the origin of the universe as a baseline. In addition there are two voices from the past.

In knowing that he knows, man finds a oneness as well as an apartness from the rest of the universe. In a sense man is an alien. The ultimate product of evolution now controls evolution. The path to truth taken by science converges with the different path taken by poets, mystics and ancient religions.

Meister Eckhart, a German Dominican theologian and mystic (c.1260 - c.1327), expressed the same idea 650 years ago:

> Creation is the Revelation, the Home and the Temple of God. Every creature is the Word of God. (Quoted in Fox, 1980, p. 57)

Eckhart acknowledges the oneness of all creation and his words are in perfect harmony with what evolution and genetic evidence tell us so clearly. He also expresses the apartness of man from creation because man knows he knows.

> God is closer to me than I am to myself: my being depends on God's being near me and present to me. So he is also in a stone or a log of wood, only they do not know it. If the wood knew God and realised how close he is to it as the highest angel does, it would be as blessed as the highest angel. And so man is more blessed than a stone or a piece of wood because he is aware of God and knows how close God is to him. And I am the more blessed, the more I realise this, and I am the less blessed, the less I know this. I am not blessed because God is in me and is near me and because I possess Him, but because I am aware of how close He is to me, and that I *know* God (italics supplied). (Quoted in Woods, 1986, p. 62)

Eckhart attracted an enormous following throughout Germany during his lifetime. In 1329, about a year after his death, he was condemned as a heretic. During his lifetime he was easily able to rebut the charges of those who conspired against him. One of these charges, apparent even in the two previous quotes, was that he downplayed original sin.

In 1980 the Order of Preachers, founded by St. Dominic, began official proceedings to have the case against Eckhart reversed. Eckhart's works are a major source of inspiration for those intent on developing a creation theology and spirituality relevant for today. Richard Woods says it thus:

> Perhaps the most pressing reason for another look at the "Eckhart Case" concerns just those crises in the Church

today which at first glance make the whole notion of a rehearing seem preposterously backward-looking. For many of these challenges and dilemmas are surprisingly like those of Eckhart's own times – a massive loss of confidence in the Church as an institution, a sense that God is remote or, indeed, "dead," a sometimes overwhelming awareness of social violence and corruption against which the message of the Nazarene stands pale and ineffective, a proliferation of alluring cults and mean-tempered sectarianism, a pervasive fear of the future and, hardly least, a vast and consuming hunger for meaning, value and love on the part of people everywhere. In the fourteenth century Eckhart's preaching and teaching revitalized the faith of much of Europe. That same mystical spirit, freed from the lurking suspicion of un-orthodoxy, may well help to dispel the spiritual confusion of our time, springing up into the parched lands of trust and friendship like a fountain in the wilderness. (Woods, 1986, p.10)

In the twentieth century few have been such a source of inspiration as the priest scientist Pierre Teilhard de Chardin (1881 - 1955). Teilhard stated unequivocally that

Man is the end-point which gives meaning to every thing that went before. To put it in a striking phrase which Teilhard borrowed from Julian Huxley, *"Man is evolution becoming conscious of itself ."* (Fernando, 1983, p. 83)

Teilhard has been of significant influence in modern religious, philosophical, economic and scientific thinking. His synthesis of evolution and Christianity at a time when the mention of the word evolution was anathema resulted in his being banned by Rome and exiled from France to America. The rejection to which Teilhard was subjected in his lifetime continues to the present day, particularly from many of those Catholic clergy who were educated in seminaries where evolution was equated with atheism. This is probably the major reason why Catholic fundamentalists ally themselves with creation science both in Australia and in America.

Catholic fundamentalism and creationism share the same mindset but there are differences. Catholic fundamentalists seem trapped in a kind of time warp. Inside is absolute truth while outside all is heresy. The limits of truth are defined by the 1545 Council of Trent at one end and the 1950 encyclical *Humani Generis* at the other.

Decisions, Decisions

The effects of creationism may be most clearly seen in the classroom but they also penetrate the whole of our society. They threaten present and future generations. The rise of creationism is symptomatic and paralleled by a decline in the relevance of Christianity. It is aided and abetted by a lack of leadership from Christian churches in the past. For this the whole Church must bear part of the responsibility. Protestant theologian Langdon Gilkey sees the loss of the traditional Catholic sense of the supernatural as a root cause.

> For on the creative resolution of this contemporary challenge to Catholicism depends the health of the whole church in the immediate future. (McBrien, 1980, p. 15, quoting Gilkey)

Raymond Brown, the eminent Catholic biblical scholar, holds similar views to Gilkey.

> The greatest danger facing religion is not the danger of new ideas; it is *the danger of no ideas at all*. Too often those who have the official task of proclaiming the word of God see no danger if people are passively content in their religion. They see a greater danger when people are restive about what they hear and when they ask themselves and others challenging questions. (Brown, 1985, p. 143)

Elsewhere Brown states:

> Religious people think they know what God wants. If one suggests to such people that it is necessary to change one's mind about what God wants in order to hear the word of God, then the offence of the Gospel becomes clear. We remember that Jesus had few problems with sinners; they

seem to have been relatively open to his message. His greatest problem was with religious people who knew already what God wanted and were therefore offended by hearing a different message from Jesus ... The preacher who asks people to change their minds is often the preacher who will be castigated for disturbing the people, precisely because it is not sufficiently stressed that Jesus was a disturbing figure and that his message presented faithfully will inevitably disturb. That disturbance does not touch simply sinful behavior but *also wrong conceptions of God's set of values and wrong understanding of doctrines.*
I see this as a particular problem in the experience of Roman Catholics. We have been emphatic that we have a set of answers and that those need to be repeated and passed on from generation to generation (italics supplied). (Brown, 1985, p. 142)

The writings of both Meister Eckhart and Pierre Teilhard de Chardin figure prominently in emerging theologies of creation based on the fact of evolution. Both writers are still officially frowned upon.

It is not too much to say that in the area of environment, pollution, nuclear wastes and so on – all those areas which threaten life on earth – Rome is silent because of a fear of embracing evolution. In present circumstances, saying that Catholics *may* believe in evolution, that evolution is a theory only, is not only evidence that the mentality of the Augustinian original sin lingers on in Rome but a clear indication of failure to read the signs of the times concerning what is perhaps the major issue confronting human beings today.

If Rome judges, or is seen to judge, the threat to the world itself as irrelevant, then it cannot complain if the world regards Rome as irrelevant. The rest of the world will fight on regardless.

Chapter 16

Causes and Conclusions

In our present age for the first time in history the resources exist which would ensure that no person in the world need lack food and shelter. This has been largely the achievement of Science and its bedfellow Technology. The global tragedy is that the reverse is the case. Millions starve to death and many more sleep hungry each night.

What is more, the biosphere itself is so under siege that higher forms of life are threatened with extinction. Who could have foreseen this even fifty years ago? What has gone wrong?

The pessimistic pre-millennial wing of creationism proclaims the end times and looks forward to them with near glee. The optimistic post-millennial wing sees these troubled times as a grand opportunity to take over the world and declare it Christian. The rest of us must search through the causes with a hope that therein may lie solutions. Not that anyone seems to have found them. However, it is in times of rapid change such as the present that the phenomenon of fundamentalism most strongly makes its presence felt.

The Displacement of Religion

The late nineteenth century and the early years of the twentieth century were times of great optimism. Science seemed on the verge of achieving a full understanding of the universe and in theory the problems of the world could be solved. World War I, which witnessed the killing fields of Somme and Ypres, put a massive dent in this optimism. Yet the myth persisted that science could do anything. The prestige and adulation which in previous ages had been lavished upon the priests and pontiffs of religion were now given to scientists. And gratefully accepted. The scientific way of knowing became predominant and was seen as the only path to truth. Art, the humanities and religion were pushed aside as inferior modes of attaining knowledge. Langdon Gilkey, professor of theology at the University of Chicago Divinity School, comments that

> where a certain form of knowing becomes so predominant as to represent for many the exclusive mode of knowing, as has frequently been the case with empirical science in modern culture, then whatever that form of knowing "knows" will represent the sole relevant and actual aspect of reality. In this way, scientific truth becomes the normative and, in fact, the exclusive form of truth, and scientific explanations represent the total and solely relevant explanations of whatever is. As we have seen, it is out of this error, which can be called the error of "scientism" or positivism, an error characteristic of an advanced scientific culture and shared in different ways by both sides, that this controversy about evolution has been generated. (Gilkey, 1985, p. 178)

But the past two decades have brought disillusion to a society confident that science would solve all problems. No longer is there the illusion that science itself is an entity quite independent of the society and culture in which it exists.

Science is dependent on governments for its funding and direction. Nowadays scientists are increasingly aware of this.

In Leonard Wibberley's *The Mouse that Roared*, Dr. Kokintz is a scientist who has invented a small hand-held bomb capable of blowing the planet to smithereens. Kokintz has been captured, with his bomb, by a small feudal duchy in Europe which declared war on the United States and sailed to conquer it with bows and arrows. As a prisoner of war he philosophizes to his captor on the plight of the scientist:

"You are a young man," he said, "and so you talk of high principles not from knowledge of them but through ignorance. The problem which confronts me is one with which you will never be confronted. Scientists such as I have become creatures of another world. Our very work deprives us of the normal moral values which guide the layman. No one understands what we do but ourselves. We communicate with each other in a language which can only be understood by ourselves and by no others – the language of nuclear physics. We know better than anyone else the terrible potentialities of our work. Yet we are bound by the laws of other men. We are gods or devils, whichever the others make us. *The harm or good which comes from our work is their choice and their doing.* It is they who decide whether millions will die or whether they will live such lives as have never been lived before – longer lives, happier lives, lives freer of disease. *Do not condemn the scientist, young man. Condemn rather the laymen of all nations who control the scientists*; the laymen who cannot agree among themselves and as a result compel us to play the part of destroyer. War existed before science. The crime which is done now is that war has made a tool and slave of science, and man's knowledge, painfully and laboriously compiled, is made the instrument of man's destruction." (Wibberley, 1981, pp. 138 - 9)

Kokintz presents a view with which many would concur. But he also indicates the effects which have led to the public image that science can solve all problems as well as scientific knowledge being the only real knowledge.

The dominating scientific community as such is as much a symptom or an effect of the power of science in modern culture as it is its cause. In effect, the scientific community has *received* its position of dominance rather than *caused* it; it has received it from the quite understandable veneration with which science is regarded by modern populations. Nevertheless, there can be little question that the community of scientists, and not least, the community of educators in science, bear some responsibility for the "scientism" which, as we have seen, has been a major cause of this controversy about evolution. (Gilkey, 1985, p. 182)

Whatever may be the final judgment, the dethronement of science from its image as a hero savior of the world has left a vacuum into which have gushed strange religious cults and sects along with witchcraft, astrology, superstitions, the paranormal and a host of other entities of which creationism is by no means the least. There is no evidence that the major Christian churches are to any significant degree filling the gap left by the demise of the myth of science in the West. They seem to be undergoing a crisis of relevancy. More people are turning to the religions of the East rather than what is perceived as the tired religion of the West. In many places the clergy and the congregation seem destined to grow old together as the proportion of young people attending church diminishes year by year. This does not necessarily mean that religion as a phenomenon is declining but that many more young adults each year perceive the established churches as simply not catering to their needs.

The decline of the scientific myth has occurred in tandem with a failure of leadership in Christianity. Together they fertilize the ground on which creationism flourishes – an age of uncertainty and swift change.

Creation Science as Science

This book has examined key areas of science and shown that in each of these areas scientific creationism simply does not measure up. It has also been shown that the whole spectrum of creationism falls if evolution is admitted as fact.

Flood geology is claimed by Henry Morris to be the linchpin of creation science. It is to creationists what evolution is to modern science. Consider just one more example, the Himalayas, a mountain chain in which Everest has pride of place. These comprise fossil-bearing sediments with the sediments angled at around 30 degrees to the horizontal. Creation geology from its first principles states that the fossils, the sediments and the 8-kilometer (5-mile) uplift were the result of the Flood 4000 years ago. This means that the sediments and fossils were laid down horizontally during the Flood and that the uplift occurred 2000 or 3000 years ago. Such an uplift would have devastated the Northern Hemisphere. The planet would still be shaking.

In contrast, geologists have shown that the Himalayas, including Mount Everest, are but the top layer of sediments 50 kilometers (31 miles) thick. These sediments were laid down during the Jurassic and Cretaceous periods between 210 million and 150 million years ago. They were subsequently uplifted very slowly as two continental plates ground together. There is no evidence of volcanic activity, which would indicate a fast uplift. The Himalayas are mostly shales and limestones deposited at an average rate of 7 centimeters (3 inches) per 1000 years. They are still being uplifted a few centimeters a year (Plimer, 1988b).

The nonsensical positions which creationists are forced to assume and defend are a result of their reading the Bible as a scientific text. Nor is there any way of escape for them except to brazen it out or invoke a miracle from their god of the gaps.

The Constants

The universal constants at this stage of our knowledge are like the fingerprints of the universe. "The universe is constructed thus" say the constants. Creationist attempts to slow down the speed of light indicate contempt for the way the world is made. They magnify the error when in their pseudo-recantation they cower in a corner and say it is a mystery known only to God. For those who believe that the Spirit of God is not confined within the bounds of any religious institution but spills over into the whole world, including science, such a position makes a mockery of a Creator.

The Laws of Nature

The laws of nature are based on observations. Thousands of scientists have painstakingly measured and recorded directly the ways in which creation runs. Bit by bit, more light has been shed on the complex ways in which the laws interlock with each other. The limits at which the laws break down provide an opportunity to move apart the veil of mystery a little more.

Creationists assert that the first law of thermodynamics was made on the seventh day of creation when God rested and that the second law was made fifty years later when Adam and Eve were driven out of the Garden. This is an affront to the Bible and is infantile science.

The Heavens

Optical astronomy can reach out to galaxies farther than 2 million light-years from earth. This is potent evidence that the universe is at least 2 million years old. Although ending in debacle, creationist attempts to slow down the speed of light were made in order to protect their basic premise that an age for the earth of 6000 years, as derived from the Bible, must be a maximum. This may be a logical approach if nothing else, but creationist founder Henry Morris finds much more in the heavens than stars. He glimpses "occasionally in Scripture" that there is a continuing cosmic warfare between good and bad angels taking place among the stars. "Then war broke out in heaven. Michael and his angels fought against the dragon" (Revelation 12:7). Planets, asteroids and the moon being closest to earth were nearest the combat zone. Hence the scars and fractures on the moon are likely remnants of this battle. Morris further claims it is likely that UFOs indicate the battle is still being waged. Such sightings indicate that the rulers of darkness are "becoming increasingly imaginative in their battles for the minds of men" (Morris, 1972, pp. 66 - 7).

Such is the astronomy of creation science.

The Crust on Which We Tread

The lithosphere is like the skin on an orange. It is the layer of solid rock which coats the earth. Beneath it, all is plastic. The lithosphere is in a state of constant change as molten material from below seeps up through some of the places where the crustal plates meet. The "ring of fire" which stretches around the Pacific

Ocean is one of the best-known examples of where the crustal plates grind together, thus producing volcanic activity.

The lithosphere is thinnest under the oceans, while mountain ranges are supported by a much greater thickness which projects into the mantle below. Unlike an orange the skin of the planet varies in thickness but like an orange it is not totally rigid and immovable. The skin which floats on the plastic mantle is to a certain degree capable of bending and deformation if the process occurs over geological time, that is, over tens or hundreds of millions of years. If the process were to occur over hundreds or thousands of years, as creationists claim, then chunks of crusts the size of continents would continually be plunged back into the mantle, there to be melted down and eventually recycled.

Moreover, as the high parts of the crust are relentlessly eroded by wind and weather, the mountains themselves are moved grain by grain to another position. The mountains, worn down to their roots, are now found as sediments. The sediments slowly sink because of the new weight on the crust. This causes other areas to be pushed up in compensation. This is an additional mode of mountain forming besides the pushing up of the crust where the tectonic plates meet. All this and most other geological processes take place over the vast eons of geological time. Geologists have over more than a century built up a detailed picture of how this and other geologic processes occurred through the sequence of fossils trapped in sediments deposited since the earliest life-forms more than 3 billion years ago. Modern radiometric dating allows the actual age of rocks to be measured. This verifies the ages of the crustal rocks calculated from sedimentation and the fossil record.

To anyone with even the slightest knowledge of geology, to believe that all this has taken place over the past 4000 years during one great flood is incomprehensible. Creation scientists don't merely tamper with geology. They totally destroy it.

Races and Memories

In 1805 Admiral Horatio Nelson decisively defeated the French in the Battle of Trafalgar. For creationists, 1805 marks the 4000th anniversary of the Tower of Babel, before which time all mankind lived together and spoke the same language. It was just 150 years after the Flood and Noah was still alive. As the senior family member he was doubtless able to regale his descendants (the whole human race) with stories of that fateful year spent upon the waters. Thus say creationists.

A harmless, even humorous story perhaps, but for creationists the evidence is as plain as the nose on your face. Australian creationists claim to have done in-depth studies of Aborigines, their culture and their stories. They also claim to have "one of the most extensive collections of Aboriginal myths and legends in the world." Their researches have proved conclusively that the Dreamtime stories are but racial memories of the Flood and the Tower of Babel. They seek financial support to carry on the good work so that Aborigines can read them before it is too late (*Ex Nihilo Technical Journal*, 1986, p. 155).

In the space of 200 years the white man has taken away the land which Aborigines had occupied for over 40,000 years. In one stroke creationists have deprived them of their religion. But creationists see themselves as impartial. They say the same about the religions of all other indigenous peoples around the globe. After all, they all dispersed to their various countries when God pushed over the Tower in that fateful year of 2195 B.C.

The moral is that creation science simply ignores, as if they never existed, all of the many disciplines, scientific and other, which are concerned with the origin of cultures and peoples.

Evolution

A litany of scientific nonsense can be taken just so far. It is 130 years since the man who started it all made the comment below. It is as appropriate today as it was then. Here is the last sentence in Charles Darwin's *Origin of Species:*

> There is a grandeur in this view of life, with its several powers originally breathed by the Creator into a few forms or into one; and that, whilst this planet has gone cycling on according to the fixed law of gravity, from so simple a beginning, endless forms most beautiful and most wonderful have been, and are being evolved. (Darwin, 1859, p. 560)

To those who have read thus far any comment upon creationism as science must surely be superfluous.

Creationism as Religion

Creationist literal interpretation of Genesis has false religious implications. Earth is a wayside stopover. People were placed on the planet by a creator who knew they would disobey and most would go to eternal damnation. Creationist use of apocalyptic writings to show that a nuclear war will soon destroy the world are scare tactics as well as a total misinterpretation of the Bible. Their image of Christ as the avenging arm of God with a nuclear warhead instead of a shepherd's staff in his hand is offensive to most Christians.

Coupled with this, the virtual deletion of the Gospels from the New Testament and the Prophets from the Old Testament is Christianity in name only. In its 2000-year history Christianity has had more than a few periods on which it can look back with shame. In these the image of Christ in the Gospels somehow came through. Creationism emphasizes Satan more than Christ. This leads to a perverted Christianity bordering on religious pornography.

Corrupter of Minds and Maker of Atheists

The admitted major function of the science of creationism is to engage the mind and thereby evangelize. By their very nature and as a basic premise, creationist materials teach that all the rest of the world is wrong. Such indoctrination in schools is brainwashing which marks a child for life. It breeds intolerance and prejudice of the worst kind, producing minds which remain closed for life.

Just Who Is a Creationist?

Creationism comprises a spectrum of beliefs differing in degree rather than kind. A strict, literal reading of a book as complex as the Bible must inevitably lead to nearly as many interpretations as there are readers. At different times and in different places particular demagogues will hold sway over others. A new variation arises when others are convinced by new emphases and priorities of interpretation by a particular individual. But at any time the same individual may be converted to another mode of creationism since the literal Bible is the factor common to all of them. In the selection process what is emphasized is just as important as what is left out. The sixty-six books which comprise the Old and New Testaments are by many authors on many topics written hundreds of years apart. There is an overall balance among all the books. Random selection can be disastrous. There is an old saying that "Any fool can quote the Bible to his own damnation."

On the other hand there are a great many Christians who call themselves fundamentalists in the sense that for them the Bible is sufficient. They feel no great need for churches or institutions. Most of these fundamentalists would find abhorrent the practices of the fundamentalism described in this book. Nor would they use deceit, guilt or threats of hell to others, nor be involved with using scientific readings of the Bible to boost their faith. Their faith is in what they feel is the experiential presence of God in their lives. Such people, and thankfully they may yet outnumber fundamental creationists, believe in a reality rather than a self-made image of God.

A Menace to Society

At one end of the creationist spectrum are those who believe the world is at the stage where Christians must take over governments and declare democracy to be contrary to the Bible. They will work for a theocratic state governed by their perception of the laws laid down in the Bible. God will ordain leadership on those Christians best suited (and we all know who they are). They will enforce the one true religion which will take over the world for Christ. History and science will be rewritten to agree with the Bible. God help the rest of us.

These are the extreme post-millennial creationists. Their campaign has already started. If it sounds familiar it may be because of resemblances to the Bible-based beliefs of the ruling class in South Africa who perceive themselves as Israel and South Africa as the promised land given to them by God. Nor can systems which maintain they own all of truth get by without inquisitions or secret police. Burning heretics and imprisoning dissidents are two sides of the same coin.

The other end of the creationist spectrum is more comfortable with material wealth which, like the contemporaries of Job, they view as a sign of God's favor. Entry to heaven is by being "born again" in one way or another. The conquistadores of old baptized the hapless natives before beheading them so they would go straight to heaven. These particular evangelists use any means to justify their end of saving souls no matter whether the souls want to be saved or not. Creation science is one of these means. It is a sign of righteousness, though incidental, that in the process the dollars flow in. In all their appeals they use many techniques to garner money for the "Lord's work." Never once in all their hundreds of thousands of exhortations have they mentioned the starving or dying in Africa, or anywhere else. May one assume that this is because such unfortunates have not been born again and thus it would be a waste of time because they are doomed to hellfire anyway.

Conclusion

For creation science, nonsense is an asset. This is because a slight acquaintance with creationism causes most people initially to dismiss it as of little consequence, mildly amusing or even harmless. But the scientific aspect of creationism is the thin edge of a wedge which forces a choice between evolution and atheism, between science and faith. To present the minds of the young with such a choice is intolerable. To set up this false dilemma using fear and guilt is morally reprehensible. To practice this on schoolchildren is downright criminal.

Creationism across its whole spectrum stands or falls on the fact of evolution. Evolution is just as much a scientific fact as the earth orbiting the sun. But the proof of a sun-centered universe goes against the senses. It is not obvious. Whereas the proof of evolution is all around us for anyone with eyes to see. Understanding the details of how the universe has evolved reaches into many of the most fascinating areas of science and develops a wonder and appreciation for the world around us.

Creationism can exist only in a Western culture in which for more than 1500 years the Christian interpretation of the Bible has become embedded as the only theory of origins, in which there has been, with a few exceptions, a failure of church leadership at every level to listen to and to communicate to their followers who have a right to know. This lack of courage coupled with silence is a further asset to creationism. As time goes on, failure to respond becomes more obvious and the silence becomes deafening. Leaders who bury their heads in the sand become unaware of their loss of respect among those who once respected them.

Till now scientists have been at the forefront in combating that aspect of creationism known as creation science or scientific creationism, and it will continus thus. But it should be obvious by now that it is not the task of science alone. Debating with creationists according to their rules and within the limits they set is usually counterproductive. It is apparent that input from theology and the humanities is needed and rightly expected. These areas are just as much under threat. Creationism has been successful in building and maintaining the myth that the dispute is between themselves and science. It is in the non-science disciplines that creationism is most vulnerable and they know it. It is arguments from these areas which they fear most.

Furthermore, it is at the grass-roots school level that creationists concentrate their efforts. It is here they have been most successful. There is a dearth of resources to combat the mountain of creationist materials and the number of personnel which they have trained in strategies for implementing creationism in schools. In Australia, unlike the United States, we do not have a national center to involve ordinary people in combating creationism in their own region. In the United States such a center exists with at least one committee in each of the states as well as in Canada. Such an organization would include not just academics but the many members of the public who would wish to join their efforts in combating creationism at the grass-roots level and in educating others. Such an organization would have under its umbrella all those with a common interest in combating creation science: the ordinary homemaker, student or man in the street together with scientists and academics from other disciplines.

Finally, this book has covered those areas relevant to an understanding of creationism. There are many books which pursue each of the different facets of creationism in much greater depth than is possible in a single volume. But the light of understanding is only the first step.

If the information in this book is effective in helping even one parent or one child it will have been worth the effort.

References

Aardsma, G.E. (1988). "Has the Speed of Light Decayed?", *Impact*, no. 179 (El Cajon, CA: Institute for Creation Research).

Alszeghy, Z. and Flick, M. (1967). "An Evolutionary View of Original Sin," *Theological Digest*, vol. XV, no. 3, autumn, pp. 197 - 214.

Anderson, B.W. (1967). *Creation Versus Chaos* (New York: Association Press).

Archer, M. (1987a). "Creationism: No Place in the Science Class," *Australian Natural History*, vol. 22, no. 4.

Archer, M. (1987b). "Squaring Off Against Evolution: The Creationist Challenge" in Selkirk, D.R. and Burrows, F.J. (eds), *Confronting Creationism: Defending Darwin* (Sydney: NSW University Press).

Arnold, P.M. (1987). "The Rise of Catholic Fundamentalism," *America*, April 11.

Asimov, I. (ed.) *et al.* (1973). *Concepts in Physics* (Del Mar, CA: CRM Books).

Baum, R. (1982). "Science Confronts Creationist Assault," *Chemical Engineering News*, 18 January, p. 24.

Beale, B. (1988). "Scientists Strike Back," *Sydney Morning Herald*, 25 June, p. 74.

Bible, Good News English Version. (New York: American Bible Society).

Born, M. (1969). *Physics in My Generation* (New York: Longman Springer-Verlag).

Bridgstock, M. (1985). "Ten Checks Upon Creation Science," *Australian Science Teachers Journal*, vol. 30, no. 4, March.

Bridgstock, M. (1987). "What Is the Creation Science Foundation Ltd?" in Bridgstock, M. and Smith, K. (eds), *"Creationism", An Australian Perspective*, 3rd ed. rev. (Sydney: Australian Skeptics), pp. 81 - 6.

Brown, R.E. (1985). *Biblical Exegesis and Church Doctrine* (New York: Paulist).

Catholic Weekly (1987). "Pecking Away at Darwin," 15 April, p. 6.

Chagas, C. (1979). "Einstein the Man," *Einstein, Galileo* (Vatican: Libreria Editrice Vaticana).

Christiansen, K. (1984). "The Natural Sciences Know Nothing of Evolution" in Weinberg, S. (ed.), *Reviews of Thirty-one Creationist Books* (Syonett, NY: National Center for Science Education).

Chure, D.J. (1984). "Walk the Dinosaur Trail: Book 1, Trails Beginning" in Weinberg, S. (ed.), *Reviews of Thirty-one Creationist Books* (Syonett, NY: National Center for Science Education).

Ciraolo, M. and Knobbe, A. (1982). "'Creationist' to Speak at U.C.," *Daily Californian*, 9 April.

Clancy, E. (1968). "Creation Beliefs in the Ancient Orient" (A4.4) and "The Hebrew Attitude to Creation" (A4.5), seminar papers published in Price, B. (1986a), *Genesis, Evolution, and Creationism* (Sydney: Catholic Education Office).

Cole, J.R. (1981). "Misquoted Scientists Respond," *Creation/Evolution* , vol. 6, no. 4, pp. 34 - 44.

Conrad, E.C. (1981). "Tripping Over a Trilobite: A Study of the Meister Tracks," *Creation/Evolution* , vol. 6, no. 4, pp. 30 - 3.

Darwin, C. (1859, reprinted 1958). *Origin of Species* (London: Oxford University Press).

Davies, P. (1983). *God and the New Physics* (Harmondsworth: Penguin).

Dawkins, R. (1986). *The Blind Watchmaker* (New York: Penguin).

Desmond, A.J. (1976). *The Hot Blooded Dinosaurs: A Revolution in Paleontology* (New York: Dial Press).

Duquoc, C. (1978 - 79). "New Approaches to Original Sin," *Cross Currents*, vol. XXVIII, pp. 189 - 200.

Eliade, M. (1967). *From Primitives to Zen* (London: Collins).

Fernando, M. (1983). "Pierre Teilhard de Chardin: An Outline of His Thought on the History and Future of Mankind" in Zonnewald,

L. and Muller, R. (eds), *The Desire to be Human* (Wassenar, Holland: Mirananda).

Finger, M. (1988). Personal Correspondence to B. Price, 28 February.

Flick, M. (1965). "Theological Problems in 'Hominization'," *Theology Digest* , vol. 13, no. 2, summer, pp. 122 - 6.

Fox, M. (1980). *Breakthrough, Meister Eckhart's Creation Spirituality in New Translation* (New York: Doubleday).

Frye, R.M. (ed.) (1983). *Is God a Creationist?* (New York: Charles Scribner's Sons).

Gilkey, L. (1985). *Creationism on Trial* (Minneapolis: Winston Press).

Gish, D.T. (1977). *Dinosaurs, Those Terrible Lizards* (San Diego: Creation-Life).

Gish, D.T. (1986). *Evolution? The Fossils Say No* (San Diego: Creation-Life).

Godfrey, L.R. (1981). "An Analysis of the Creationist Film, Footprints in Stone," *Creation/Evolution* , vol. 6, no. 4, pp. 23 - 30.

Gosse, P.H. (1857). *Omphalos: An Attempt to Untie the Geological Knot* (London: Van Voorts).

Grigg, R. (1988). "From Fish to Gish," *Creation ex Nihilo*, vol. 10, no. 2, March - May.

Haight, R. (1985). *An Interpretation of Liberation Theology*, (New York: Paulist).

Ham, K. (1988). "God's Reminders – Sin," *Creation ex Nihilo*, vol. 10, no. 2, March - May.

Hardin, G. (1984). "Marketing Deception as Truth" in Montague, A. (ed.), *Science and Creationism* (London: Oxford University Press).

Hawking, S.W. (1988). *A Brief History of Time* (London: Bantam).

John Paul II (1986). *Man, the Image of God, Is a Spiritual and Corporeal Being* (General Audience: 16 April, n. 7).

Jukes, T.H. (1982). Memorandum to Department Chairpersons, University of California, Berkeley, 6 April.

Jukes, T.H. (1984). "The Creationist Challenge to Science," *Nature*, vol. 308, March, pp. 398 - 400.

Jukes, T.H. (1986a). Letter to the Editor, *Nature*, vol. 321, June, p. 722.

Jukes, T.H. (1986b). "U.S. Supreme Court to Review Louisiana

Appeal," *Nature*, vol. 324, December, pp. 423 - 4.

Jukes, T.H. (1988). Personal Correspondence to B. Price, 10 July.

KDF (1988). "Paul Ellwanger Strikes Again," *Creation/Evolution*, vol. 8, no. 1, January - February.

Klayman, R.A. et al. (1986). *Amicus Curiae Brief of 72 Nobel Laureates, 17 State Academies of Science, and 7 Other Scientific Organizations, in Support of Appellees* (Washington: Caplin and Drysdale, Chartered Accountants).

Knobbe, A. (1982). "Gish: 'There Must Be a God'," *Daily Californian*, 12 April.

Lexicon Universal Encyclopedia (1987). (New York: Lexicon).

Marty, M.E. (1983). "Burdened Schoolmasters," *Christian Century*, vol. 100, 28 September, p. 86. He quotes from Martin, R.E. (1983), "Reviewing and Correcting Encyclopedias," *Christian School Builder*, vol. 15, no. 9, April, pp. 205 - 7.

McBrien, R.P. (1980). *Catholicism* (Minneapolis: Winston Press).

McIver, T. (1988). "Christian Reconstructionism, Post-Millennialism and Creationism," *Creation/Evolution*, vol. 8, no. 1, January - February.

Moore, R.A. (1983). "The Impossible Voyage of Noah's Ark," *Creation/Evolution*, vol. XI, winter.

Morris, H.M. (1972). *The Remarkable Birth of Planet Earth* (Minneapolis: Bethany).

Morris, H.M. (ed.) (1974). *Scientific Creationism* (El Cajon, CA: Master Books).

Morris, H.M. (1977). *The Beginning of the World* (Denver: Accent).

Morris, H.M. (1978). *The Twilight of Evolution* (Grand Rapids, MI: Baker Book House).

Morris, J.D. (1986). Letter to the Editor, *Nature*, vol. 321, June, p. 722.

Moyer, W.A. (1985). "How Texas Rewrote Your Textbooks," *Science Teacher*, vol. 52, no. 1, January, p. 22.

Newspapers:

The Daily Californian (Livermore, CA).

The Herald (Livermore, CA).

The Independent (Livermore, CA).

The Valley Times (Livermore, CA).

Nogar, R.J. (1965). "The Wisdom of Evolution," *Theology Digest*, vol. 13, no. 4, winter, pp. 269 - 80.

Origins, Film Series Handbook (1983). (Illinois: Films for Christ Association).

Osgood, A.J.M. (1984). "The Times of the Judges – A Chronology," *Ex Nihilo Technical Journal,* vol. 1, pp. 141 - 58.

Osgood, A.J.M. (1986a). "A Better Model for the Stone Age," *Ex Nihilo Technical Journal* , vol. 2, pp. 88 - 102.

Osgood, A.J.M. (1986b). "The Times of Abraham," *Ex Nihilo Technical Journal* , vol. 2, pp. 77 - 87.

Osgood, A.J.M. (1987). "Rock Hard Orange!" *Creation ex Nihilo,* vol. 10, no. 1, p. 11.

Overton, W.R. (1981). "Creationism in Schools: The Decision in McLean versus the Arkansas Board of Education," published verbatim in *Science* (1982), vol. 215, February, pp. 934 - 43.

Parker, G.E. (1979). *Dry Bones ... and Other Fossils* (San Diego: Creation-Life).

Plimer, I. (1986). "Creation Science – The Work of the Devil," *Australian Geologist*, Newsletter 61, December.

Plimer, I. (1988a). Personal Correspondence to B. Price, 19 February.

Plimer, I. (1988b). Personal Correspondence to B. Price, 15 August.

Price, B. (1986a). *Genesis, Evolution, and Creationism* (Sydney: Catholic Education Office).

Price, B. (1986b). *The Two Books of God* (Sydney: Catholic Education Office).

Price, B. (1987). *The Bumbling, Stumbling, Crumbling Theory of Creation Science* (Sydney: Catholic Education Office).

Prigogine, I. and Stengers, I. (1984). *Order Out of Chaos* (London: Heinemann).

Rifkin, J. (1985). *Declaration of a Heretic* (Melbourne: RFK).

Ritchie, A. (1987). "Testimony of the Rocks" in Selkirk, D.R. and Burrows, F.J. (eds), *Confronting Creationism: Defending Darwin* (Sydney: NSW University Press).

Science (1982). 29 January, p. 486.

Scott, E.C. and Cole, H.P. (1985). "The Elusive Scientific Basis of

Creation Science," *Quarterly Review of Biolology*, vol. 60, no. 1, March, pp. 21 - 30.

Selkirk, D.R. and Burrows, F.J. (eds) (1987). *Confronting Creationism: Defending Darwin* (Sydney: NSW University Press).

Setterfield, B. (1984a). "C-decay and the Red-shift," *Ex Nihilo Technical Journal*, vol. 1, pp. 71 - 86.

Setterfield, B. (1984b). "Geological Ages of the Earth, Planets and Flood Strata," *Ex Nihilo Technical Journal*, vol. 1, pp. 52 - 9.

Setterfield, B. (1986). "Carbon-14 Dating, Tree-Ring Dating and Speed of Light Decay," *Ex Nihilo Technical Journal*, vol. 2, pp. 169 - 88.

Skehan, J.W. (1986). *Modern Science and the Book of Genesis* (Washington: National Science Teachers Association).

Snelling, A. (ed.) (1984). *Ex Nihilo Technical Journal*, vol. 1.

Snelling, A. (ed.) (1986). *Ex Nihilo Technical Journal*, vol. 2.

Stockton, E., Nanson, G. and Young, R. (1987). "Chronology and Palaeo-environment of the Cranebrook Terrace (Near Sydney) Containing Artifacts More Than 40,000 Years Old," *Archaeology in Oceania*, vol. 22, no. 2, pp. 72 - 8.

Taylor, P. (1987). "Dinosaur Mania and Our Children," *Creation ex Nihilo*, vol. 9, no. 3, June, pp. 23 - 8.

Wheeler, A.G. (1987). "Creation Science in Australia," *Quadrant*, March, pp. 58 - 62.

Whitcomb, J.C., Jr. (1972). *The Early Earth* (Grand Rapids, MI: Baker Book House).

Whitcomb, J.C., Jr. and Morris, H.M. (1961). *The Genesis Flood* (Grand Rapids, MI: Baker Book House).

Wibberley, L. (1981). *The Mouse That Roared* (New York: Bantam).

Wieland, C. (1987). "Scientific Bombshell," *Creation ex Nihilo*, vol. 10, no. 1, pp. 12 - 14.

Wilder-Smith, A.E. (1977). *The Natural Sciences Know Nothing of Evolution* (San Diego: Creation-Life).

Willey, B. (1959). "How the Scientific Revolution of the Seventeenth Century Affected Other Branches of Thought" in *Reader 1, The Project Physics Course* (Sydney: Horowitz).

Woods, R. (1986). *Eckhart's Way* (Wilmington, DEL: Glazer).

Zhou Ming Zen and Li Yan Xian (1983). "Teilhard's Contribution to Science in China" in Zonnewald, L. and Muller, R. (eds), *The Desire to be Human* (Wassenar, Holland: Mirananda).

Zindler, F.R. (1985a). "The Case of Big Daddy," *American Atheist*, May.

Zindler, F.R. (1985b). "The Case of Big Daddy II," *American Atheist*, July.

Zindler, F.R. (1986). "Maculate Deception: The 'Science' of Creationism," *American Atheist*, March.

Zindler, F.R. (1986 - 87). "Report from the Center of the Universe," *American Atheist*, sampler.